Copyright © 2017 Sang Lee

All rights reserved. No part of this book may be used or reproduced in any manner whatsoever without written permission of the author.

Printed in the United States of America.

First Edition published on August 7th, 2017

ISBN-13: 978-1974314027

ISBN-10: 1974314022

TABLE OF CONTENTS

Introduction 7

1. Cell Phone 8
2. Nail Strip 9
3. Crocs Shoes 10
4. Barbed Wire 11
5. Post-it 12
6. Nylon String 13
7. LEGO 14
8. Zipper 15
9. Cotton Swab 16
10. Choco Pie 17
11. Triangular Underwear 18
12. Mechanical Pencil 19
13. Heelys Roller Shoes 20
14. Flexible Straw 21
15. Keurig Coffeemaker 22
16. Miracle Mop 23
17. Windshield Wipers 24
18. Pay at the Pump 25
19. Handheld Calculator 26
20. Coca Cola 27
21. Screwdriver with Light 28

22. Windowed Envelope 29
23. Sugar Cube 30
24. Pillow Pets 31
25. Inline Skates 32
26. Sony Walkman 33
27. Correction Fluid and Tape 34
28. Drone 35
29. Selfie Stick 36
30. Bread-Slicing Machine 37
31. Thermometer and Thermostat 38
32. Viagra 39
33. Ivory Soap 40
34. Prius 41
35. Tesla 42
36. Shampoo 43
37. Videotape 44
38. Fitbit 45
39. Uniqlo HEATTECH 46
40. Rubik's Cube 47
41. Minecraft 48
42. WhatsApp 49
43. Dyson Vacuum Cleaner 50
44. Rice Cooker 51
45. Nintendo Wii 52
46. Blue Jeans 53
47. Water Bottle with Compass 54
48. Toothpaste DIspenser 55
49. Corning Ware 56

iii

50. Smartphone 57
51. Sweetener 58
52. Penicillin 59
53. Anaesthesia 60
54. Rubber Tire 61
55. Copy Machine 62
56. Electric Cart 63
57. Sim Card 64
58. Rabies Shot 65
59. Ramen Noodles 66
60. Microwave Oven 67
61. LED Light 68
62. Amazone One-Click Shopping 69
63. 3D Printer 70
64. Mickey Mouse 71
65. GoPro Camera 72
66. Overwatch Game 73
67. Razor Shaver 74
68. Shaved Ice Dessert (Bingsoo) 75
69. Pokemon Go 76
70. Two-Way Electric Socket 77
71. Cancer Diagnosis Kit 78
72. Shake Shack Burger 79
73. Uber 80
74. Online Unversity 81
75. Metro Subway 82
76. Air Conditioner 83
77. Sunglasses 84

iv

78. Hula Hoop 85

79. Java Software 86

80. Cooking App 87

81. Ringer's Solution 88

82. Sovaldi Tablets 89

83. Panera Bread 90

84. Sulwhasoo Cosmetics 91

85. Artificial Heart 92

86. Self-Driving Car 93

87. Artificial Intelligence 94

88. Yamaha Clavinova 95

89. Apple Pencil 96

90. 3D Galaxy Gear 97

91. Fuji Cosmetics 98

92. Microsoft Office 99

93. Probiotics 100

94. Fidget Toys 101

95. Canned Food 102

96. Rain-X 103

97. 3D Printer for Food 104

98. Perfume 105

99. Hangul (Korean Alphabets) 106

100. Robot Vacuum and Dog Walker 107

Conclusion 109

The thing that hath been, it is that which shall be; and that which is done
is that which shall be done:
and there is no new thing under the sun.
Is there any thing whereof it may be said, See, this is new?
It hath been already of old time, which was before us.

ECCLESIASTES 1:9-10
KING JAMES VERSION (KJV) BIBLE

HOW TO BECOME A MILLIONAIRE BY INVENTION?

100 Stories of Successful Inventions

INTRODUCTION

How can you become rich and successful?

How can you become a millionaire by invention?

It seems that great inventions and ideas are what caused the success of many others in the past! Would you like to meet the Stories of these inventions from now on?

Let's read their invention stories, learn from those inventors and help you to reach financial success!

We are not trying to know what 100 famous inventions are. Actually, there are thousands of famous inventions and it is impossible to know all of them and we cannot wait to be a millionaire for a long time. Rather we are trying to learn the secrets behind 100 famous inventions.

The secrets of their inventions are actually quite simple: most great inventors tried to make their lives or their loved ones' lives easier and refused to accept everything for granted. Instead, they believed in themselves and in the fact that they could change the world by finding a better way to complete a task.

You will be surprised to find that it is not difficult to become an inventor into a millionaire. For instance, some inventors created new things with flash of genius, but most inventors improved the current invention by modifying the current invention, adding a new function to the existing invention, combining two inventions into one or using the current invention for another purpose.

Once we know the secrets behind 100 famous inventions, we might duplicate their success Stories and become an inventor and a millionaire.

Let's begin the journey!

HOW TO BECOME A MILLIONAIRE BY INVENTION?   1st Story / Total 100 Stories

100 Stories of Successful Inventions

Cell Phone

First of all, we will discuss an invention that seems to have become a necessity of life: the cell or smart phones. The smart phones have altered the aspects of everyday life just about everywhere. The smart phones first started as cell or mobile phones.

In 1964, Joel Engel of Bell Labs provided the primitive cellular telecommunication service to the Detroit police department.

In 1973, Dr. Martin Cooper, called the "father of the cell phone" invented the first handheld cellular phone and led the team that brought it to market. Dr. Cooper's first cell phone weighed a hefty 850 grams, much larger than the modern Samsung Galaxy 7 that weighs approximately 150 grams. Dr. Cooper said a wireless communication device used by Captain Kirk in a Star Trek movie inspired his cell phone development. After more than a decade of hard work, Motorola released the first commercial mobile phone "brick" Dynatac in 1983. With a charging period of 10 hours, people were able to talk for 20 minutes. (Note: If you charge a phone for about an hour now, it can run for roughly10 hours).

NTT's system, the first cellular telecommunications system, was installed in 1979 in Tokyo, Japan.

The first test of Advanced Mobile Phone System (AMPS) was conducted at a location between Washington, DC, and Baltimore in 1981 and the US Federal Communications Commission (FCC) finally certified commercial cellular communications services.

AMERITECH opened America's first analog cellular service (AMPS: Advanced Mobile Phone Service) in Chicago in 1983.

In 1984, Korea Mobile Communications (now SK Telecom) also launched mobile phone service and mobile phone service is now commonly seen worldwide.

The idea for the cell phone came from Star Trek movie.

HOW TO BECOME A MILLIONAIRE BY INVENTION? 2nd Story / Total 100 Stories

100 Stories of Successful Inventions

Manicure Strip

People are always trying to save time and money. If you have some ideas how to save time or money, the ideas might lead to a great invention.

On April 1987, a 29-year-old young man, Park Hwa-young, was riding a bus from 32nd Street in Manhattan, New York, to 86th Street. He was a graduate of Hanyang University College of Music and studied in the United States. As he was on his way to a vocal lesson, a woman in the back of the taxi wore nail polish on her fingernails. When the car stopped, Park thought. "If I can eliminate the time it takes for nail polish to dry, I will be very popular."

Now, Park Hwa-young is the president of the company, INCOCO, which currently sells millions of dollars of Manicure Strips. INCOCO is the world's only 100% dry manicure maker and distributor. Since its establishment in 1988, it has revolutionized cosmetics and is the first cosmetic company to transform 100% nail polish into dry manicure instead of bottle. INCOCO is now headquartered in New Jersey, USA. American products are being developed and manufactured by the US headquarters for packaging, mass production for cosmetic brands around the world. INCOCO has been pioneering the 'Drying Manicure' market with its patented technology. In addition to the convenient application of nail polish, INCOCO Dry Manicure also offers women a wide variety of shades.

The busy people of today's world have welcomed the dry manicure because INCOCO's invention has saved their time necessary for painting and drying nail polish.

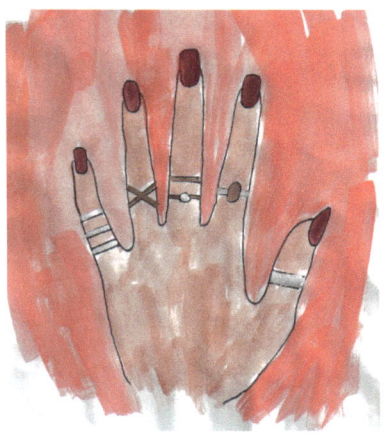

The nail strip saves time for modern women.

HOW TO BECOME A MILLIONAIRE BY INVENTION?

3rd Story / Total 100 Stories

100 Stories of Successful Inventions

Crocs Shoes

President George W. Bush and other Hollywood stars wore Crocs shoes, which boosted their popularity throughout the world, as people rushed to joint the trend.

What are Crocs shoes? Who made them? What are their secrets?

'Crocs shoes', a kind of rubber footwear, are products that dominate the summer shoes market each year. CROCS shoes, which first appeared in Colorado, USA in 2002, are popular worldwide in the United States, Europe, and Japan.

Crocs shoes are popular due to its lightweight and comfortability. Though it may seem rubbery, it is not smooth or slippery due to its 'croslite' material. According to the company's explanation, it is said that Crocs have the effect of reducing muscle fatigue by more than 60% compared to standing barefoot because of its patented ergonomic design.

Also, Crocs shoes have many holes for ventilation, which allow feet to breathe and the shoes to dry quickly.

Crocs has expanded its customer base with all season shoes brands while adding design innovations such as combining styles with sophisticated designs based on comfort.

Crocs Shoes, made of lighter and more comfortable material, became popular fashion items and began as little ideas and grew into one of the greatest of inventions that have changed the world.

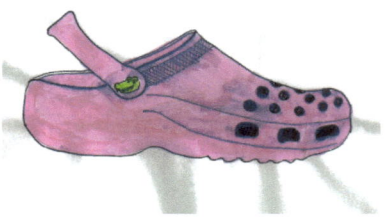

Comfortable Crocs shoes

HOW TO BECOME A MILLIONAIRE BY INVENTION?   4th Story / Total 100 Stories

100 Stories of Successful Inventions

Barbed Wire

A barbed wire fence commonly has a number of uses in agriculture, but it can also be used for other reasons: containment, protection, division and deterrents. Barbed wire is formed simply by twisting hard pieces of wire together to form razor sharp points at various places.

It is a cheaper and quicker alternative to building a large and elaborate wooden or stone fencing structure.

In 1826, Joseph Glidden, a 13-year-old shepherd, was trying to keep his flock of sheep from fleeing over the fence. While the sheep were grazing, he was reading his book relaxed and dreaming about his future. Then one afternoon, he heard the voice of his boss urgently yelling. "Joseph! What the hell are you doing? Look at that." Joseph, who saw what his master pointed at, was astonished. A few sheep had crossed the fence and already ruined the crops. At that time, the fence was made with some wires over wooden spikes. Glidden then went to the fence and kept watch over the sheep, but the sheep kept trying to spoil the crops of his neighbors when he wasn't looking. Joseph began to struggle day and night thinking, "What better way is there?" Then to Joseph's amazement, he noticed that the sheep were avoiding the roses with the thorns nearby the fence. From that day on, Joseph cut small pieces of rose thorns and tied them to the fence. One day, Joseph discovered another amazing fact: he saw a twisted wire and then a wire spike on the cut. At the moment Joseph's mind came up with brilliant thoughts. Right! He would go straight to the blacksmith to get the pliers and wires, make the wire spike pieces, and twist wire spike pieces over the wire to put on the fence. The finished wire was much more durable, and the edges were sharp. Joseph filed a patent application with the help of the rancher, and the barbed wire quickly became popular.

Joseph, who had a positive attitude toward new possibilities was able to invent barbed wire, and of course, became richer when he found discomfort in everyday life and he did not miss the opportunity to make a change for the better.

The idea of spiked wire came from rose thorns.

HOW TO BECOME A MILLIONAIRE BY INVENTION?   5th Story / Total 100 Stories

100 Stories of Successful Inventions

Post-It

The POST-IT is an invention that is now commonly seen in offices and schools around the world.

It all began in the mind of a man named Arthur Fry on Sunday in 1973, when he stood up to sing at the Sunday Choir.
He dropped the bookmark from his book, the one that he had put in to know which page to pick up for the next exercise. So then he thought it would be nice to use a weak-glue made by his colleague, Spencer Silver, a scientist at 3M Company who loosely placed weak-adhesives for the easy sticking and removing of paper. The next day, he went straight to Silver, to get his weak-glue sample and start experimenting.
It was in that moment, that the history of 3M Post-it began.

More than 30 years from then, you can see this sticky note wherever you go in the world because this product has been sold by more than billions of dollars so far.

Fry, who used a method of simple observation and questioning, noticed that the piece of scrap he put in the book fell out, and did not miss the opportunity to become a great inventor by solving his own problem and became rich by making people's life easier.

Fry invented Post-It by using a method of observation and questioning.

HOW TO BECOME A MILLIONAIRE BY INVENTION?   6th Story / Total 100 Stories

100 Stories of Successful Inventions

Nylon String

"It is thinner than a spider web, yet a yarn that is tougher than a wire!"

One morning in February 1937, people from all over the world had received the paper and read an article that introduced an amazing miracle fiber called nylon. It brought a great revolution in the 20th century clothing culture and was created by a young chemist named Wallace Carothers.

In 1927, Carothers started working for the DuPont lab, an American textile company. Soon Carothers became the head of the Research Department the following year, due to his outstanding research achievements. He also helped Dupont succeed in producing Dupren, a synthetic rubber, for the first time. Carothers then announced that he invented about 40 new substances.

In 1930, a fellow researcher named Dr Hill, who was looking into the microscope, alerted Carothers. "Doctor, look at this!" he said. "Isn't this a thread-like compound?" The two were extremely excited about the new discovery due to the expectation that artificial fiber might be born, a substance better than rayon. Carothers, who has been involved in research for the following five years, invented nylon in 1935 by combining it with air, water, and coal. The original name of the nylon is 'Polma 66', meaning that the number of carbon atoms in the compound was 66.

Finally, on May 15, 1940, nylon stockings that were lighter than silk were sold at department stores throughout the United States. They were twice as expensive as silk stockings; yet more than five million pairs were sold on the first day.

Many great inventions are often found by chance. But the inventors who are always working and who are ready for improvement only pick up the accidental inventions. Recognizing opportunity when it knocks requires a sense of open-mindedness, an accurate judgment, and a keen sense of determination.

Nylon string, thin and tough, was invented by accident.

HOW TO BECOME A MILLIONAIRE BY INVENTION?   7th Story / Total 100 Stories

100 Stories of Successful Inventions

Lego

Lego, which has been praised for being the "toy revolution", is one of the greatest toys in human history.

A carpenter named Ole Kirk Christiansen, born in Denmark in 1891, created Lego. In his youth, he showed curiosity toward everything, and was unique among the other children. During adulthood, Christiansen often received the envy of his fellow carpenters, and he always had a project to work on in the home building business. By the way, Christiansen did not forget to analyze the merits and demerits of making a model before building the house for his client; as we can see, he was a very meticulous man. At that time, carpenters could not have imagined the big picture, but this practice of making a model made him even more famous. Christiansen's modeling continued, and in the meantime, he became involved in making block-shaped toys. In the end, he stopped his career as a carpenter and set up a toy company. He named the toy as well as his company "Lego" which translated to "having fun." in Danish. Christiansen, with innate dexterity and endless effort, continued working day and night in the pursuit of the highest quality toy in his lab. As a result, the world's first Lego was finally born in the form of a block. Lego's popularity among children was explosive around the world. In the early days, blocks were created by cutting wood, but in the 1940s it became possible to mass-produce using plastics by Christensen's son, Godtfred. In 1963 Godtfred introduced the ten basic criteria of the Lego system:

1. Functionality of play is infinite
2. For both boys and girls
3. Suitable for children of all ages
4. Durable throughout the year
5. The health and comfort of children
6. Suitable for moderate play time
7. Able to increase development, fantasy, and creativity
8. Amplify the value of more play
9. Able to be supplemented easily
10. Quality must be perfect.

Lego was originally intended for children up to 12, but there were also adults who fell back to playing with Lego blocks during various occasions such as playing with children. Of course, inventions enjoyed by both children and adults naturally become big hits.

A carpenter, who used to make miniature models for houses, invented Lego toy.

HOW TO BECOME A MILLIONAIRE BY INVENTION?

8th Story / Total 100 Stories

100 Stories of Successful Inventions

Zipper

What is often a part of your clothes, yet not a button, a ring, a garment, a bag, a purse, or a glove? It's a zipper! Let's find out this fabulous invention!

When a man named Whitcomb L Judgson went to work, he noticed that there were 12 buttons on the Walker boots on his feet, but decided that it was so tedious to tighten them one by one. So through trial and error he made a continuum of metal rings. The earliest zipper resembled a claw that is closed when the metal tab was pulled up and is opened when the tab was pulled down. Its usage was similar to today's zipper, but the problem was that the material was iron, which rusted and became difficult to open after one wash. Salespeople had to teach customers how to use it, not to mention the problem of its intricate shape, which complicated mass production and became too expensive, putting the manufacturing company in a great crisis.

At this time, there was an engineer named Gideon Sundback who emigrated from Sweden to the US. Judson recognized this man's skills immediately and hired him with twice the average salary, and Gideon later became Judson's son-in-law. Sundback devised a new way by locking the teeth of the zipper together. After about four years of research in 1905, the material was changed from iron to copper alloy, which did not rust. In 1917, he produced the zipper with modern copper alloy.

Later, a coin purse manufacturing company bought the zippers to put on the coin purse. Goodrich (now a tire company) also began placing zippers on its rubber boot products. During World War II, the company supplied war materials and expanded with more than 5,000 employees and Sundback became rich. Later on, Yoshida Tatao in Japan developed a new plastic zipper that overcame disadvantages of metallic zippers on the conventional military uniforms and he produced plastic zippers based on cheap labor in Japan. The company is now YKK, which accounts for 60% of the global zipper market.

Judgson has invented a zipper, but the zipper has been improved continuously with the persistence and tenacity of several inventors. This is a small tool that opens and closes quickly, but it has saved time for people and has become a great invention.

The first zipper was invented to replace 12 button shoes to save time.

HOW TO BECOME A MILLIONAIRE BY INVENTION? 9th Story / Total 100 Stories

100 Stories of Successful Inventions

Cotton Swab

When we look around us at everyday household items, the many of these items were once big hit inventions in the past. Many of the world's hit products were born from the idea of necessity, as "necessity is the mother of invention". The invention of the cotton swab is an example of this.

An inventor named Leo Gerstenzang observed his wife as she wiped the ear of a child with cotton, and felt that it could be quite dangerous. This was because earwax was typically wiped with ear picks, and in some cases, cotton or cloth was loosely attached to the tip to remove moisture from the ear after bathing. It is not known exactly when the ear picks began to be made, but it is presumed that they have been used for a very long time since these objects have been found in old tombs. When the cotton swab was finally created, it was originally made only to extract earwax, but gradually began to be used in other situations.

Although the cotton swab was invented with a hint from the ear picks, the swab and ear picks are separate objects and are still used as separate objects.

There is no perfect invention in the world, so we will always wait for continuous improvement, and that improvement will most likely become another invention.

What is the secret to Gerstenzang's success? He believed that ear picks are made of metal, which can be dangerous for your ears when you remove earwax. "Even if you attach cotton or cloth to the metal end, you can still injure your eardrum if the debris of cotton or cloth is separated from the metal and the metal touches the inside of your ear. Then you can remove the moisture from your ears, and you can get dirt on the swab with the moisture in your ears."

In 1923, Gerstenzang established Leo Gerstenzang Infant Novelty Co., and produced cotton swabs under the trademark Baby Gays. The cotton swap was a big hit and he became a millionaire. In 1926, the product name was changed to Q-TIPS BABY GAYS. Where Q is quality. Since then, Q-TIPS cotton swap became one of essential household items in our lives.

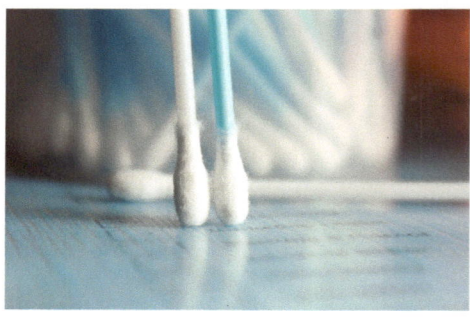

Cotton swap was invented by combining a small wad of cotton and a short rod.

HOW TO BECOME A MILLIONAIRE BY INVENTION?   10th Story / Total 100 Stories

100 Stories of Successful Inventions

Choco-Pie

Choco Pie is a popular snack among soldiers serving in the military, as well as among students and children in South Korea. According to Orion, the sales of Choco-Pie exceeded 100 billion won (US$92.04 million) in global sales in the first quarter of 2015, reaching a record high. It is made of a soft marshmallow filling between two round soft biscuits coated with chocolate on the outside.

One similar dessert was born in 1917 in Chattanooga, Tennessee, USA, in the form of biscuits containing marsh mallows. The name of the Choco Pie was 'Moon Pie' because its round shape resembled the full moon. Moon pies are still sold in stores in the United States, but now Koreans as well as people in the global community are more familiar with Orion's Choco Pie, because Orion's Choco Pie has become a best seller worldwide.

So who first introduced Korean Choco Pie to the world? In 1973, researchers from Oriental Confectionery (now Orion) were looking for new confections, and came across Choco Pie in a foreign cafe. Astonished by the delicious taste, they saw potential for the dessert to become a hit, and went to work by creating cookies with chocolate at as soon as they arrived home.

One researcher at Orion said, "It seemed like I could succeed at the beginning of my research, but I failed many times. It was not easy to find a way to make biscuits, marshmallows and chocolates blended together."

A year later, Orion finally succeeded in its research. The secret was in the moisture of the marshmallow. With little moisture, marshmallow turned out dry, hard, and easily cracked. However, with enough moisture, the biscuits were made smoother, and a thin layer of chocolate was also added to sweeten the pie. Orion introduced this new sweet pie to the world under the name Choco Pie, and it quickly became very popular. However, Orion could not properly manage the trademark of 'Choco Pie' at the time, so other confectionery companies produced chocolate pies under the same name. However, some of the loyal fans, including the author, who have tasted Orion Choco-Pie, still insist on buying it rather than its similar brands.

Choco-Pie, invented by Orion Company, is selling like hot cakes around the world.

HOW TO BECOME A MILLIONAIRE BY INVENTION? 11th Story / Total 100 Stories

100 Stories of Successful Inventions

Triangular Underwear

Panties were originally designed to cover the entire lower half of the female torso, but have become increasingly shorter and smaller over time.

The triangular underwear was patented in Japan in 1951 after its invention by a woman named Takako Sakurai. On a hot summer day, she saw her grandchildren wearing underwear that fell down to their knees. At that time, it was uncomfortable to wear the underwear, which closely resembled short pants, and the underwear became uncomfortable in the hot summer without air circulation. One thought came to Mrs. Sakurai, who was sick of seeing her grandson with uncomfortable underwear: the underwear simply needed to cover a minimal amount of skin. Sakurai cut an old sack made of cotton cloth, and made a triangular panty by sewing a hole for the legs. From that moment, lightweight and convenient underwear was born. Sakurai received a patent for this underwear, and soon people started to buy them. The love for her grandchildren made Sakurai create the simple underwear that quickly became a big hit in 1954. She was a peculiar inventor who created only a panty collection, including a 'triangle pant' called a micro panty, a unique panty with a narrow stitch, a thigh panty with stockings, and a baby diaper cover. Then she thought, "Inventions are so easy. I want to invent a few more." The short panty series that came out immediately caught the public eye. Until then, panty makers used to put a cloth over the hips to emphasize the difference between the waist and hip curves. However, Mrs. Sakurai discarded this step altogether. Her design allowed for a snug fit, and the wearer didn't need worry about the sewing thread breaking because it was made of one piece.

Another Sakurai's inventions: Thigh panties made from one cloth and i Atom panties for infants became popular, too. Sakurai's panties were mass-produced by Toyo Rayon, one of Japan's leading clothing companies, and later swept the whole world. Sakurai was given a royalty of 300,000 yen per year, which was considered a large sum of money at the time.

Sakurai, who did not accept uncomfortable underwear for granted, improved the underwear by herself and thus became a great inventor in the process.

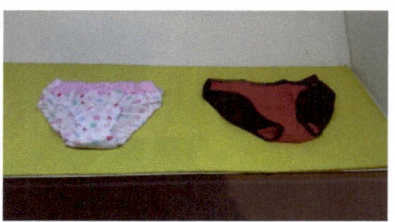

Mrs. Sakurai's triangular underwear was invented for her grandson.

HOW TO BECOME A MILLIONAIRE BY INVENTION? 12th Story / Total 100 Stories

100 Stories of Successful Inventions

Mechanical Pencil

The mechanical pencil, which has been loved by many students around the world, is an invention that was created by necessity to save time.

Its inventor, Hong Lai, was a Taiwanese blacksmith who grew up learning his skills alongside his father. Lai worked day and night at the laboratory barn. With the extraordinary effort and tenacity, he produced over a hundred works, but he was always being chased because he could not hit one of them properly. He had to write many ideas or research steps on the 20-page notebook by pencil overnight. Therefore, one of the most troublesome tasks was sharpening his pencil. Was there any way he could write comfortably without having to sharpen his pencil? Tired of the inconvenience, he stopped all other research and investigated methods for creating a pencil that did not require sharpening. He started thinking that if he could control the pencil lead, the problem would be solved. But it was not easy. He grappled with ideas in his mind and felt sick for a month until his flash of genius comes. As he grabbed the toothpaste before brushing his teeth, he thought, "Aha! Let's apply the principle of pressure used with a tube of toothpaste to the pencil." Within a few days, the mechanical pencil was born by putting the pencil lead in a cartridge that would be inserted in a hollow plastic pipe. By 1972, the pencil was devised to remove the cartridge and push the pipe rim to bring out another piece of pencil lead once the lead ran out by pressure.

The first person to come up with the patent registration was the president of BaekJeung Co., a leading stationery company in Taiwan. Hong Lai sold the patent for around $200,000, a high amount that was unprecedented in Taiwan. The sale soon became a hot topic in the business world and the president of BaekJeung Co. was scorned as a madman. However, with the start of production, the things have changed. Even though they made the products day and night at the factory, they could not produce enough products for purchase orders. Inevitably, the factory started to grow exponentially and the mechanical pencils became popular worldwide.

Indeed, Hong Lai refused to accept the traditional belief that pencils should always be sharpened and invented a mechanical pencil and became a rich man.

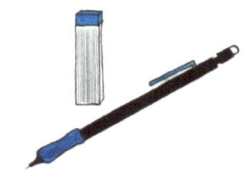

Mechanical Pencil with automatic lead loading made peoples' lives easier.

HOW TO BECOME A MILLIONAIRE BY INVENTION? 13th Story / Total 100 Stories

100 Stories of Successful Inventions

Heelys Roller Shoes

Many children play around on the street wearing shoes with wheels,

One time, I was even surprised to see dazzling lights on a child's pair of shoes passing by me at night.

Who created these odd shoes?

In 1998, Roger Adams, who would later become the president of Heelys, was recovering from a recent bankruptcy. Adams loved roller-skating and when Adams was only nine months old, he was the 'youngest roller skater' who was listed in the Guinness Book of World Records.

He said, "What if I have shoes with which I can walk or skate at the same time?" Heelys shoes look just like normal sneakers, but wheels are hidden in the soles or can be removed. With these wheels, people can skate around quickly instead of just walking.

Adams patented the shows in 1999 and started selling the shoes.

Heelys shoes have become popular among children and teenagers, and the shoes are popular among even white-collar workers who want to speed up their work. Adams, who was once the president of a bankrupt company, is now a millionaire because of his brilliant ideas.

However, it may be important to note that experts warn riders to keep the balance carefully with protection on their knees and elbows to prevent injuries.

Heelys Roller Shoes with which people can walk or skate at the same time.

Shoes with Wheels and Flashing lights

HOW TO BECOME A MILLIONAIRE BY INVENTION? 14th Story / Total 100 Stories

100 Stories of Successful Inventions

Flexible Straw

One time my father-in-law, who was 80 years old at the time, asked me to buy flexible straws for him. Out of curiosity, I researched these straws and found that old or sick people favored these straws because they were able to drink milk or other beverages with flexible straws quite easily. Who do we have to thank for this?

In 1936, an American man named Joseph B. Friedman noticed his young daughter, Judith, trying to drink a milkshake out of a straight straw while sitting in his sweet shop in San Francisco. He figured that it would be a good idea for her to drink from a straw that would bend, and soon the flexible straw was born. Friedman applied for a patent in 1937 and then applied for five more patents both in the US and abroad for the shape and manufacturing method of flexible wrinkle straw in 1950. When the flexible straw was first invented, it was indeed extremely popular in hospitals since patients who could only lie down were able to drink with ease.

As Friedman's invention has proven, the love for family exerts great power. The creation of flexible straw is proof that concern for loved ones can truly lead to the greatest inventions, as it was Friedman's love for his daughter that led to the invention.

Friedman's flexible straw that was submitted to the U.S. Patent Office

A straw that can bend or flex and make that familiar upside-down L shape was invented for the inventor's daughter.

HOW TO BECOME A MILLIONAIRE BY INVENTION? 15th Story / Total 100 Stories

100 Stories of Successful Inventions

Keurig Coffee

When I am alone in my office, I use Keurig coffee instead of using 10-cup coffee maker. Keurig coffee is a single serve coffee brewer. The coffee is brewed directly into a cup. Keurig coffee lets everyone make his or her favorite kind of coffee for each cup very quickly without worrying about dishwashing later.

One spring day in 1995, John Silva felt dizzy and his heart started racing, and he was rushed to the emergency room. After his diagnosis, it turned out that the cause of his issues had been coffee addiction. Silva had been drinking 30 to 40 cups of coffee a day. For the past three years he and his business partner, Peter Dragon, had been inventing a new coffee machine. The name of the coffee machine they built was the "Keurig", which means "excellence" in Danish. This became the start of a coffee machine company that now raises $ 800 million in sales a year.

There is a growing interest in coffee as more people drink coffee even just to get up in the morning. The secret to the popularity of the Keurig machine is in K-CUP, short for "Keurig cup", which is a disposable plastic cup-shaped capsule in which the coffee is sealed. Once the K-CUP is inserted in the upper part of the coffee machine, one must select the amount of water, and the water injector creates a hole the aluminum lid of the cup to allow hot water to filter through the coffee. When the coffee is made, the capsule is separated from the coffee machine and directed to the trashcan and there is no need for dishwashing. Today, for busy people and especially coffee addicts, Keurig coffee makes delicious coffee in a short time. This time saving coffee machine for one person while making a fresh coffee became a great invention.

The Keurig coffee machine and various types of K-cups.

HOW TO BECOME A MILLIONAIRE BY INVENTION?   16th Story / Total 100 Stories

100 Stories of Successful Inventions

Miracle Mop

"Success is moving from failure to failure without loss of enthusiasm," said Winston Churchill. Many great inventions are made at the worst moment for the inventor.

One day, Joy Mangano, a divorced mother with a distant ex-husband, an old grandmother and three children to care of, broke a glass cup of wine at home. Suddenly, an idea came to Joy's head as she mopped up the broken glass. It seemed dangerous to hand-squeeze the mop now filled with glass pieces. That is the moment when she decided to invent the revolutionary mop called the 'Miracle Mop', with a head made from a continuous loop of 300 feet (90 meters) of cotton that can be easily wrung out without getting the user's hands wet. However, inventing a great product is just beginning. Mangano had to sell her great product somehow. While Mangano contacted many companies and many investors, they persistently rejected her invention. Finally, one day she succeeded in selling the 'Miracle Mop' in 20 minutes in the home shopping broadcast, and it became the best-hit product in the history of home shopping. However, Joy, who had gained an immense profit, did not stop after this one success.

When she saw a coat falling from her hanger, she deciede she wanted a coat hanger that did not allow her clothes to fall to the ground. She changed the shape of the hanger so that it did not

Slope downwards, and soon this invention too became a hit product. The story of Joy Mangano's success inspired many people and it was made into a movie called JOY in 2015.

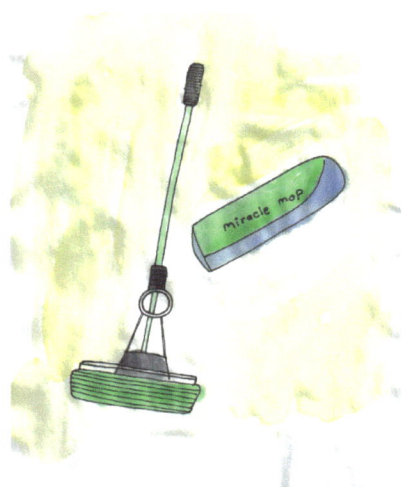

Miracle Mop was invented by accident when Joy, the inventor, broke a glass cup of wine.

HOW TO BECOME A MILLIONAIRE BY INVENTION? 17th Story / Total 100 Stories

100 Stories of Successful Inventions

Windshield Wiper

Without windshield wipers, it becomes difficult to see through a window covered with snow, rain, or dust. A windshield wiper generally consists of a metal arm, pivoting at one end with a long rubber blade attached to the other. The arm is powered by an electric motor although pneumatic power is also used in some vehicles. The blade is swung back and forth over the glass, pushing water or other forms of precipitation from its surface. Most automobiles use two synchronized radial-type arms, while many commercial vehicles use one or more pantograph arms.

The inventor of this handy tool was an ordinary housewife by the name of Marie Anderson, living in Birmingham, Alabama, USA. However, since the wipers were manual, it was rather impractical to use.

In 1967, Professor Kearns invented and patented a time-controlled wiper that was inspired by the idea that a person's eyes tear up more frequently when foreign objects like dust enters the eyes. Ford and Chrysler, however, ignored Kearns' patent and commercialized a time-controlled wiper.

As a consequence, Ford paid a $ 10 million fee for patent infringement to Professor Kearns. Kearns then proceeded also to file a lawsuit against Chrysler to receive $18.7 million compensation. However, he missed the statute of limitations for other automobile companies and was unable to collect any more compensation.

"Flash of Genius" film describes Professor Kearns as a brilliant inventor and a very persistent person in the lawsuits with Ford and Chrysler.

The movement of human eyes inspired the idea of Time Controlled Windshield wipers.

HOW TO BECOME A MILLIONAIRE BY INVENTION? 18th Story / Total 100 Stories

100 Stories of Successful Inventions

Pay at the Pump

All gas stations in New Jersey and Oregon in the United States offer only "full service" and "mini service". In these states, attendants are required to pump gasoline because state laws prohibit customers from pumping with self-service. Actually, Self-service gas station has become popular in other states with the invention of Pay-at-the-pump system around 30 years ago.

In 1991, Amoco company, the world's largest oil company (now BP company), started to develop the pay-at-the pump system, with which the customer can put gas in their vehicles using a credit card without the help of the gas station's employees. The purpose of this was to reduce labor costs. The development team worked on the technology for over a year, and the final result was a big hit. Amoco and its gas stations made a lot of money by saving labor costs. Other oil companies soon followed the wave by developing similar systems, and most gas stations were well run without employing a large number of employees. This self-service system has helped machines replace people, causing employees anxiety for their future. Despite this, the automatic or self-service payment system, which started at a gas station, soon spread to other types of stores a few years later, including grocery stores, pharmacies, and other stores.

Also, artificial intelligence may soon be the ones to diagnose patients replacing doctors, and machines may replace factory workers, robots may replace soldiers, and drones may replace airplanes and helicopters. How do we prepare and teach our children about computers that are evolving and replacing people's jobs? The only way to educate the future generation is to prepare for the new world, as former Google CEO Eric Schmitt said. The upcoming generations must learn creativity, social skills, and negotiation skills, along with computer skills, leadership, and persuasion skills.

At the same time, students should be taught how to embrace computers and Artificial Intelligence while using them as collaborators and advisers.

Lastly, we should teach our future generation about the unreasonable but important things that only human beings could do: helping the poor and weak people with the compassion and serving and donating through churches, charities and social service organizations.

Pay-at-the-pump gas stations do not require gas station employee's help.

25

HOW TO BECOME A MILLIONAIRE BY INVENTION? 19th Story / Total 100 Stories

100 STORIES OF SUCCESSFUL INVENTIONS

HAND-HELD CALCULATOR

When I was a child, I learned how to use the abacus at school. Today, the abacus is no longer used, and many people carry a smart phone with a calculator app instead. So who invented the first calculator?

In 1972, a company named Casio Calculator released the Casio Mini with the sales phrase 'the world's first personal electronic device'. The typical calculator at that time was a large-scale, installable business device. The price was about 30,000 yen (about $ 300), which was too expensive for average person to buy. So the Casio Mini, which was about the size of a palm and priced at 12,800 yen (about $ 120), was a truly innovative product. The idea of such a Casio Mini came from Vice President Katsuo Yukio, CEO. He idea was born in a bowling alley, a place that was very popular among young people at the time. At that time, Yukio was often found at bowling alleys. But there was no machine that could digitally calculate the score, requiring people to calculate scores manually. So the first idea was Yukio had was that there needed to be a calculator that would be easy to use at the bowling alley. Though the calculator was very expensive then, it can cost as low as $5 nowadays.

However, Yukio was still not satisfied with its initial success and continued to develop calculators and released cheaper calculators with smaller, more computational capabilities. Soon Casio had sold more than a billion electronic calculators worldwide since 1972. When asked about the secret to keeping Casio calculator popular always, Mr. Yukio replied, "It is important to have a sense of crisis and to keep improving Casio calculator. If you are satisfied, you do not improve."

The first portable electronic calculator Casio Mini created in 1972

HOW TO BECOME A MILLIONAIRE BY INVENTION?   20th Story / Total 100 Stories

100 Stories of Successful Inventions

Coca Cola

What do you need when you are eating greasy foods like pizza, fried chickens, buttered popcorns, etc.? I would think of Coca-Cola. Who invented Coca-Cola?

In 1886, Coca-Cola was invented by a pharmacist named John Pemberton, otherwise known as "Doc." He fought in the Civil War, and at the end of the war he decided he wanted to invent something that would bring him commercial success.

Usually, everything he made failed in pharmacies. He invented many drugs, but none of them ever made any money. So, after a move to Atlanta, Pemberton decided to try his hand in the beverage market.

In his time, the soda fountain was rising in popularity as a social gathering spot. Temperance was keeping patrons out of bars, so making a soda-fountain drink just made sense.

And this was when Coca-Cola was born. However, Pemberton had no idea how to advertise. This is where Frank Robinson came in. He registered Coca-Cola's formula with the patent office, and he designed the logo. He also wrote the slogan, "The Pause That Refreshes." Coke did not do so well in its first year. And to make matters worse, Doc Pemberton died in August 1888.

After Pemberton's death, a man named Asa Griggs Candler rescued the business. In 1891, he became the sole owner of Coca-Cola. It was when Candler took over that one of the most innovative marketing techniques was invented. He hired traveling salesmen to pass out coupons for a free Coke. His goal was for people to try the drink, like it, and buy it later on. In addition to the coupons, Candler also decided to spread the word of Coca-Cola by plastering logos on calendars, posters, notebooks and bookmarks to reach customers on a large stage. It was one step in making Coca-Cola a national brand, rather than just a regional brand. A controversial move on the part of Candler was to sell Coca-Cola syrup as a patent medicine, claiming it would get rid of fatigue and headaches.

In 1898, however, Congress passed a tax in the wake of the Spanish-American war. The tax was on all medicines, so Coca-Cola wanted to be sold only as a beverage. After a court battle, Coca-Cola was no longer sold as a drug.

Now, around 2 billion servings of Coca-Cola drink products sold every day.

Coca-Cola was invented as medicine and has become the most popular beverage in the world.

HOW TO BECOME A MILLIONAIRE BY INVENTION? 21st Story / Total 100 Stories

100 Stories of Successful Inventions

Screw Driver with Light

The pyramids of Egypt, built over 2000 years ago, have various mysteries. One of them, a rather technical one, is how the Egyptians built the pyramids in the dark. Though there are remaining clues on how the pyramids were constructed, there is little evidence left to give clues to what light sources were used.

Forty years ago, the research team of Nagamori Electric Power Company of Japan had similar worries. At that time, the screwdriver was used simply to screw and unscrew. However, in many cases, it was necessary to deal with screws in dark areas of machinery. As this problem became apparent, the task of research became clear. "The driver must touch the dark corner. Is there any good way to solve this problem?" But he had no good and concrete ideas to solve the problem. The research team of Nagamori Electric Company went to the field to see field engineers' know-how's. Technicians were shouting at every corner with a flashlight. Yes, adding a flashlight to the driver was all it took. They replaced the driver's bag with clear plastic, put a battery and a little bulb in it, and then made the end of the bag into a lens shape, so that the light from the bulb would be projected intensely onto the end of the screwdriver. This new driver was sold immediately, but there was a flood of orders both at home and abroad, including from Pan Am Airlines, Japan Airlines, Scandinavian Airlines and Japan Defense Agency.

Although there is no accurate record of sales, in less than two years after the start of production that Mori Electric Company has become a midsize company from a small company without a struggle.

Screwdriver with an attached lamp

HOW TO BECOME A MILLIONAIRE BY INVENTION? 22nd Story / Total 100 Stories

100 Stories of Successful Inventions

Windowed Envelope

When you receive a bill, you can see the address and name of the recipient and the return address through the transparent cellophane.

These envelopes reduce the typing effort by half because they do not require typing the return address. So who invented the cellophane windowed envelope that saved people a little time?

A man named Americus F. Callahan of the United States, who was an ordinary salary man, came to think that a typist working in the same office one day, who had to type extra by looking at the contents of the address and name of the recipient and the same content in the envelope. "It may be necessary to type the same thing twice, but it's a big waste of manpower and really unproductive. Is there any other way?" Callahan thought very hard, but no one except Callahan thought that this issue could be improved. However, Callahan, who had a full time job, was unable to continue his research. It was just frustrating to see the duplicate typing repeated every day.

One day, Callahan stopped at a grocery store to buy a handkerchief, and found a way to solve the problem. "What color do you want?" asked the owner of the grocery store. After his selection, the owner quickly found the desired color in the wrapped paper. The secret was simple. There was a little window made of cellophane so that the handkerchief inside the wrapping could be seen. Upon returning home, Callahan looked for an envelope, cut out the recipient's address and name in a rectangle, and attached the cellophane stripped from the handkerchief wrapping over the cutout. Now he could fold his papers so that the address and the name of the recipient would appear through the cellophane. The commercial envelope with transparent cellophane was made in an hour!

Callahan succeeded in acquiring the design patent, and the windowed envelope that we commonly use today was born.

Envelopes with Return Address save time by preventing duplicate typing.

HOW TO BECOME A MILLIONAIRE BY INVENTION? 23rd Story / Total 100 Stories

100 Stories of Successful Inventions

Sugar Cube

When I was a child, I liked to eat sweet sugar cubes as candy. Who invented these sweet and convenient sugar cubes that can be so easily put into coffee or tea?

In 1958, the American Sugar Company of the United States began to research sugar cane packing because sugar is easily melted in hot temperature. There was a huge flood of ideas coming in because the Sugar Company promised to pay $ 200,000 as the reward for the solution. However, no good ideas came forward. Then one day, a man named John came to the company and brought his ideas so he could meet the boss. His clothing was so shabby that the staff thought John was a crazy person and tried to chase him away. But John was desperate and persistent. When he ran into the company, the president appeared. "What happened?" The boss asked the staff. At the moment, John quickly pulled out the bag of sugar in his pocket and put it out in front of his boss. "My boss, I've got a good idea, look closely, there's a pinhole in the package, and that's my idea." The president said that he didn't understand. But John persisted. "My boss, I'm a sailor on the deep-sea fishing boat, and we spend three months on the boat, and this packaging is the product of my company, we have been sailing for 3 months." The president became curious and asked him how the packaging worked. "There is a vent window in the storage, so it does not melt. That's the principle." Both the president and the staff nodded as if they started to understand.

The president hired John as his researcher for a salary of $200,000 thinking that the idea of this packaging would guarantee the success of his company. A mere vent hole led to John's success as well as the success of American Sugar Company that would later dominate the world.

Sugar cubes with vent holes that prevent melting

HOW TO BECOME A MILLIONAIRE BY INVENTION? 24th Story / Total 100 Stories

100 Stories of Successful Inventions

Pillow Pets

Stuffed animals are a universal toy of any child's bedroom. In fact, if a child has only one toy, it's likely to be a stuffed animal. From tigers to penguins to dogs to elephants, the fluffy toys come in all shapes and sizes to suit nearly any taste. I also heard that some ambulance people carry stuffed animals for children patients. Many people have made stuffed anmals and one person invented Pillow Pets by combining the stuffed animal with pillows.

In 2003, Jennifer Telfer was living in San Diego with her husband, Clint, and her two sons. In an interview with CNBC, Telfer states that she came up with the idea of functional stuffed toys when her oldest son, then seven, flattened out one of his stuffed animals from sleeping with it often. She decided on a combination of a pillow and a stuffed animal as the concept for Pillow Pets. Telfer and her husband proposed a meeting with a plush manufacturer at a tradeshow in Las Vegas, and decided to sell the toys themselves through their company, CJ Products. Following the meeting, the Telfers were prepared to sell the first toys.

The first Pillow Pet was "Snuggly Puppy". During Christmas season, Jennifer Telfer first sold Pillow Pets at mall kiosks, but failed to sell them all: "Every single parent would walk up, and their kids would love them. The parents would often say something along the lines of, 'No, it's just another stuffed animal." Because they didn't know that it opened up to a pillow.

A few weeks later, Telfer discovered that the Pillow Pets were very popular during the Christmas shopping season. She later ordered a container of 7,000 Pillow Pets, now in six different animal varieties, and sold all of them in three months. By the end of 2010, CJ Products earned $300,000 in gross sales. Soon, gross sales of $7,000,000 were recorded at the end of 2009. Telfer recalled: "I didn't realize that we were gonna be at this level, or people would love my idea as much as I loved it."

In 2012, Dream Lite Pillow Pets were introduced, along with Glow Pets in 2013.

Children love stuffed animals. That's why Pillow Pets are popular.

HOW TO BECOME A MILLIONAIRE BY INVENTION?   25th Story / Total 100 Stories

100 Stories of Successful Inventions

Inline Skates

Wouldn't it be nice to skate during the summer as well as during the winter? Who invented Inline skates?

John Joseph Merlin experimented with single- to many-rowed devices worn on feet in 1760.

In 1849, Louis Legrange of France invented Inline skates, designed to work like ice skates during periods of warm weather. Legrange designed the skates for an opera where a character was to appear to be skating on ice. The skates were problematic and unsuccessful as the wearer could not turn nor could they stop.

In 1979, Scott and Brennan Olson, who had pondered this questions as well, found one strange looking four-wheel roller. The boots were made poorly, but the overall design resembled ice skates. The Olsons tried splitting the skate and put the wheels on ice skate boots. Soon they realized the potential of the cross-training inline and redesigned them with plastic wheels and hockey boots, which made it possible to skate like ice skating. After several years of development and testing, Scott and Olson founded the Chicago Roller Skate Company, the parent company of Rollerblade, Inc., with the help of several individual investors in 1982, earning tens of million dollars in sales, the company becomes one of the fastest growing and profitable companies due to Scott's and Orson's curiosity and continuing research. Because of Scott and Orson's big and good question of "how can we skate all the times?" now it's now common to see people enjoying skating in hot summers as well as in cold winters.

Rollerblading skating allows for skating all year round

HOW TO BECOME A MILLIONAIRE BY INVENTION? 26th Story / Total 100 Stories

100 Stories of Successful Inventions

Sony Walkman

Only a few decades ago, everybody had to have a Sony Walkman.

The students recorded lectures, young people listened to music when they were jogging or studying, and elderly people were listened to the radio at home.

According to the memoirs of "Kuroki Yasuo", the father of Walkman, one of his researchers used to listen to music by modifying his small recorder and Yasuo started developing Walkman. However, at the same time, in 1976 Andreas Pavel, a German inventor, made a portable player similar to the existing Walkman in 1976. Pavel, who had registered a patent in Japan and in Italy won a lawsuit partially against Sony after a few years of lawsuits. Because the "Walkman" brand name is so strong many of the smaller Sony devices (except for most radio and IC recorders) are sold under the name of "Walkman". Initially, Sony released a tape player with record function, and then it released a taper player with a radio function.

Afterwards, Sony released a CD player with radio function and dominated the market for a long time.

However, Sony's portable music device market did not last long. Soon the Walkman was pushed aside by smaller MP3 players with more capacity, then by Apple's IPODs, and finally, what we often use today, by smart phones.

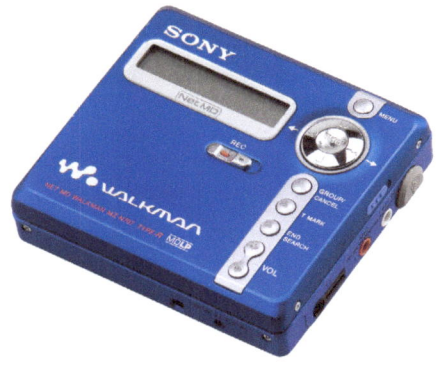

Sony's Tape and Minidisc Walkman

33

HOW TO BECOME A MILLIONAIRE BY INVENTION? 27th Story / Total 100 Stories

100 Stories of Successful Inventions

Correction Fluid and Tape

Time-travel back in the past is popular in dramas and movies these days. In the movie 'Frequency' that was released a few years ago, main characters could change the past with a ham radio, fix past mistakes and catch criminals. Our shameful past or mistakes cannot be changed, but in paper documents errors can easily be corrected with either correction fluid or tape. Also with word processor software, you can delete the mistakes on the electronic documents easily with the keyboard on your computer.

The ink correction fluid, now a necessity in the office, is an invention created by a divorced woman during a time of crisis. Betty Nesmith Graham, who was struggling to make money to raise her little son right after World War II, was working as a secretary at a bank. But her crisis began when IBM's electric typewriter came out. Electric typewriters that write letters with a slight touch produced a lot of typos compared to manual typewriters. Betty, who was unfamiliar with the batter, has been working to fix it.

One day, when she was scolded by his boss as an incompetent secretary, she came up with an idea of painting it over from her painting experience.

Afterwards, she put the white paint in a bottle and then fixed it with paint and brush. The other staff members who have come to know of her secrets have asked her to share the magic liquid.

Betty began selling white paints with 'Mist Takeout', a hand-written label, and then asked her son's chemistry teacher to assist her in developing a more powerful and quick-drying correction fluid. With her success, Betty was able to make an eventual profit of $ 47.5 million.

Correction Fluid

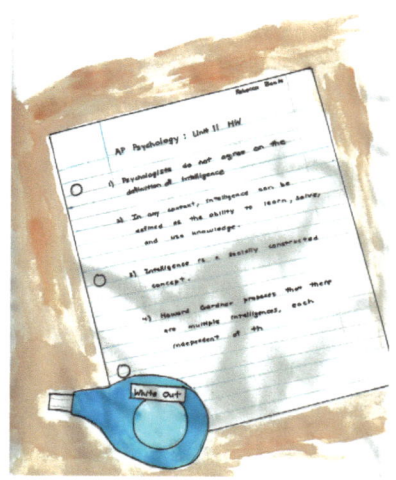

Correction tape

HOW TO BECOME A MILLIONAIRE BY INVENTION?

100 Stories of Successful Inventions

Drone

Nowadays, there are cameras for each traffic light, for a car black box, for each smart phone, and now even for a drone. But with too many cameras, we might feel like being watched constantly by Big Brother in 1984, a dystopian novel by George Orwell.

Where did the drones originate from and what will the future of drones be?

An unmanned aerial vehicle (UAV) is a craft designed to carry out a designated mission without boarding the pilot. Another name for this unmanned airplane that we commonly use is the "drone" due to the humming noise this machine produces. It often comes with various types of equipment, such as optical, infrared, radar, sensor among many others.

According to the US Department of Defense, the launch of an unmanned airplane was made in the early 1930s during the bombing of Britain in World War I, but it was not used well because of the more effective cannons and rockets.

After the 1991 Gulf War of the Middle East, it became the instrument that convinced the utility of the unmanned aircraft for military purposes. In recent years, unmanned aerial vehicles have been developed in many countries around the world.

Many military experts expect the unmanned system to rise to the core of future power. Now, research and development of unmanned aerial vehicles have become a contest for military science and technology and it is the fastest growing industry in the aerospace industry. Unmanned aerial vehicles (UAVs), which have been studied for military purposes, have been applied to the private sector in recent years. It is used not only for military purposes such as reconnaissance, electronic warfare, deceit, attack, and target, but also is used for surveillance, research and development, shooting, crime, logistics, and communication.

Civilian aircrafts are currently collecting information by observing the ground in the air and approaching where they are not accessible to collect scientific data. Drones are actively used in movies and filming, and are used in crime investigations to help arrest criminals. In the future, more companies likely to invest in drone technology, which is sure to be a big growing field for commercial and military purposes.

We see more and more drones in the sky for two reasons: helping people by flying high and remote areas and taking lives for military purpose.

HOW TO BECOME A MILLIONAIRE BY INVENTION? 29th Story / Total 100 Stories

100 Stories of Successful Inventions

Selfie Stick

One of the biggest worries when traveling with families may be asking foreigners to take family pictures.

In the 1980s, Hiroki Ueda, a Japanese developer of the camera company Minolta, came up with the idea of a selfie stick after one incident at the Louvre in Paris, France. He had asked a boy for a picture, but the boy ran away with the camera. Paranoid, Ueda was reluctant to ask for a shot during the rest of the trip, so he developed a bar that attached the camera to an extending pole. Because there was no LCD display in the film camera, he put a mirror in front of the camera to display himself. At that time, however, the camera was too heavy to attach to the pole, and the photographs that were printed were often cut off. Ueda, who held 300 patents, was not successful in selling his invention. Most people often used tripods and set a timer to run and take a group photo instead of using Ueda's selfie pole.

Later on, a man named Wayne Prom, a Canadian toy company employee, invented the selfie-stick for the second time in the early 2000s. Yet he did this without the knowledge of Ueda's work. Though he was not the inventor of the original selfie stick, he is addressed as the man who invented the present fashion selfie-stick. Prom also got the idea of the selfie-stick from the experience of waiting for a foreigner who speaks English while on vacation in Europe. Confident of commercial success, Prom promoted selfie-sticks through fairs and online shopping. However, after 10 years of his success, soon fake or counterfeits flooded and he did not earn the expected profit any more. Prom was delighted to develop something interesting and create a new culture. In addition to earning him a large profit, Prom's Selfie-stick became the best invention chosen by Time magazine in 2014.

Selfie-Stick was invented because it is not always easy to find a photographer for foreign tourists.

HOW TO BECOME A MILLIONAIRE BY INVENTION? 30th Story / Total 100 Stories

100 Stories of Successful Inventions

Bread-Slicing Machine

I recently went to a restaurant called Wildfire. The free bread, corn bread, and black bread were so delicious that I could not finish a lot of main dish ordered later on. How do they keep soft bread and keep its original taste and shape even after cutting them in slices?

Back in 1912, Otto Loewer, of Missouri, USA, suddenly sold a well-run jeweler, which sells and repairs jewelry and watches. The reason is to make a machine to slice bread. Before then, it was common to use a knife just before eating since bread would quickly become hardened when cut in advance. However, Otto believed that if he had packed the bread in advance, he would keep it in a soft, long-sleeved, easy-to-eat bowl. But his research and inventions of bread making machines were longer than planned, and finally in 1928 'pre-sliced' bread began to be sold. His breading machine and 'pre-sliced' breads were popular, and 80% of manufactured bread was pre-sliced and sold with success.

However, in 1930, Otto's competitor, Wonderbread, immediately upgraded the breading machine and started to sell breads in advance, so Otto won the US market initially with the big success of his machine, but his name was not well known. The steady efforts of so many unknown original inventors like Otto have made our lives more comfortable, even if they are not recognized today.

Otto and his Bread-slicing Machine invention made the bread easy to eat.

HOW TO BECOME A MILLIONAIRE BY INVENTION? 31st Story / Total 100 Stories

100 Stories of Successful Inventions

Thermometer and Thermostat

I used to get up in the morning and watch the weather forecast on TV.
Now I watch the weather report on my smart phone every morning in the bed. Especially in Chicago when there is a big difference in daytime and nighttime temperatures, it is easy to catch a cold if people do not check weather forecasts in advance.

Who invented the thermometer? The invention of the first thermometer dates back to the beginning of the 17th century. Galileo Galilei invented a gas thermometer and shared his joy with his disciples. The people who were with Galileo on the day are known as Santorius, De Llevel, and Prad.

However, the most significant development of thermometers was the invention of Fahrenheit (°F) and Celsius (°C) thermometers.

German physicist Fahrenheit invented the Fahrenheit thermometer in 1714 using mercury. The water temperature was set at 32 and the boiling temperature was set at 212, using the unit of degrees Fahrenheit.

In 1742, the Swiss astronomer Celsius set the temperature of water at zero and one atmospheric pressure to 100, the unit for this was called degrees Celsius.

In 1906, a man named Mark C. Honeywell produced a company called Honeywell Heating Specialty Co. to produce automatic control devices that could adjust the room temperature.

Since then, Honeywell's room thermostats are everywhere in homes, businesses, offices and factories all around the world.

With the invention of the thermometer, we can measure the temperature of a solid such as food, a liquid such as water, or a gas such as air

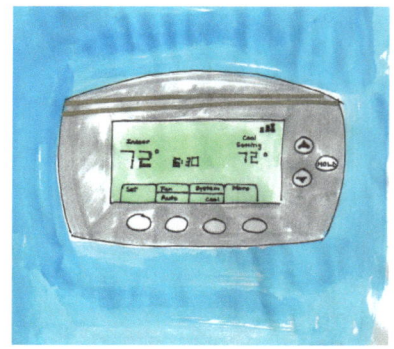

A thermostat is the main control unit for a heating or cooling system.

HOW TO BECOME A MILLIONAIRE BY INVENTION? 32nd Story / Total 100 Stories

100 Stories of Successful Inventions

Viagra

Sometimes the best inventions are discovered by accident. As English political theorist Algernon Sidney said, 'God helps those who help themselves," which emphasizes the importance of self-initiative and agency.

One of the famous inventions by accident is Pfizer's Viagra, which was surprisingly because it was originally not developed for erectile dysfunction. The original purpose of Viagra was to treat hypertension in a new way. Viagra immediately showed effectiveness and passed the process of testing with little side effects in animal experiments. It was then necessary to undergo a human clinical trial. In the first step, the drug was administered to healthy people. The results of the study showed that it was ineffective as an antihypertensive agent and that it was much less effective than nitroglycerin, which was developed long ago as a treatment for angina.

In this testing situation, the 1992 drug tolerance trial (an experiment in which maximum doses were administered to observe side effects) did reveal some interesting side effects or benefits for people with erectile dysfunction. Due to the fact that the huge amount of research funds had already been invested, the research result could not be abandoned at this point. The company actively sought to "save" the research results. From then on Pfizer began to focus its research direction on erectile dysfunction.

In May 1994, Pfizer found that 12 patients with erectile dysfunction were treated with Viagra, resulted in a positive effect in 10 patients. This news was reported to the Urology Society, and the doctor 's response was positive. Since then, Pfizer has undergone thorough clinical trials several times and finally obtained a new drug license with the brand name of Viagra from the Food and Drug Administration on March 27, 1998. The experiments could be accomplished quickly because the same ingredients were already being used in animal tests and passed the first clinical stage tests. In terms of economic feasibility of new drug development, it could not be better than this. With hundreds of millions of dollars in sales a year it has become one of Pfizer's representative drugs.

Viagra, an accidental invention, became a big hit medicine by helping some men with erectile dysfunction.

HOW TO BECOME A MILLIONAIRE BY INVENTION? 33rd Story / Total 100 Stories

100 Stories of Successful Inventions

Ivory Soap

As I have mentioned in the previous invention Stories, many of the best tools of our world have been created upon pure mistakes. Ivory Soap is one of them.

The world's largest household goods company, P & G (Procter & Gamble) is nicknamed 'Ivory Towers'. This was because P&G became famous and a global company by selling lots of Ivory Soaps. William Procter and James Gamble, co-founders of the company, used to make candles and soaps separately in the east coast and west coast. In 1837, Proctor and Gamble joined together and founded P&G Company. During the American Civil War (1861-1865), the opportunity to make money was a shining opportunity as they supplied soap and candles as war materials. In 1879, the company continued to develop innovative soaps. Until then, they had sliced heavy, sticky lumps and sold them as soap, but P & G soon developed a soap bar that floated in the water. At the time, soap was so heavy that people often lost it bathing in the rivers.

Surprisingly, this feature of floating in the water was actually unintended fortune. During the manufacturing process, the heat was mistakenly applied for too long, creating a dense air layer, which made soap with air holes in it, which would float in the water.

Advertising also played an important role in Ivory's hit. In 1882, the first newspaper ad copy was 'Ivory soap floating in water, purity 99.44%'.

Starting in 1923, P&G decided to set up an ad agency and advertised Ivory in newspapers. Ivory was the first color advertisement during TV baseball broadcasts. In the afternoon, when many housewives watched TV, P&G focused on commercials of Ivory Soap before and after radio and TV dramas and the drama with Ivory Soap commercials received the nickname of 'Soap Opera'. Ivory has been a pioneer of P & G for more than 100 years.

Procter and Gamble's original Ivory Soap, floats in the water with accidental air layers.

HOW TO BECOME A MILLIONAIRE BY INVENTION? 34th Story / Total 100 Stories

100 Stories of Successful Inventions

Toyota Prius

Would it ever be possible to create an eco-friendly, yet highly efficient car?

Toyota believed so and started developing the hybrid car Prius since 1994. Toyota launched the Prius in 1997 after 4 years of research and modification.

A hybrid means using two or more power sources together. Depending on the flow of power, the engine and the electric motor operate in parallel, and the engine produces electricity only as the electric motor is driven in series. At low speeds, the Prius rolls with only electric motors, so there is little noise and vibration. Due to lack of sound, this car can be quite dangerous both for the driver and the pedestrian, especially those who are blind or who are occupied with their phones. From the 3rd generation Prius car, the car plays the recorded engine sound to solve this problem with blind people.

In 2016, the Prius went up to more than 50 miles per gallon of gas. At first appearance, the reaction was not so good because the oil price was cheap like 15 and 20 dollars per barrel.

Later in 2000, when the oil price became expensive and went over $100, the car began to sell at rapid rates as customers scrambled to save costs from gas and the car were sold out quickly. Nowadays, the Prius, like a Tesla electric car, is plugged into a power source and can be charged overnight.

Toyota plans to further increase its investment in hybrids in the future and plans to invest nearly 20 billion dollars in annual expenditures, including research and development expenses and facility investment.

It is interesting to see the competition between Toyota's hybrid cars and other electric cars such as Tesla.

Toyota Prius, the pioneer of hybrid cars, can go up to more than 50 miles per gallon of gas.

HOW TO BECOME A MILLIONAIRE BY INVENTION? 35th Story / Total 100 Stories

100 Stories of Successful Inventions

Tesla Electric Car

Many electric car companies and other cleantech companies have failed in making electric cars. Tesla Company has succeeded in making electric car.

"325,000"!

This is the number of pre-orders received for Tesla Motors' model 3 in a few weeks. Even if Tesla sells only 325,000 cars per year, it can generate sales of $ 14 billion revenue. Model 3 is priced at $ 35,000. Rechargeable batteries are composed of more than 6,800 18650 Li-ion batteries, which are widely used in laptops. Most of the electric cars are really batteries. In recent years, Tesla has begun building the world's largest 18650-battery plant (the "Giga Factory") with Panasonic in Nevada to meet future electric vehicle sales plans and demands for its solar business (solar city). Outsourcing companies have developed eco-friendly electric vehicles as so-called affordable models, but Tesla Motors uniquely chose a niche strategy to maximize and enhance the benefits of electric vehicles to produce electric sports cars. In addition, Tesla is concentrating on establishing electric vehicle infrastructures such as charging stations in many locations in the United States. Tesla's future looks brilliant with the success of the Model 3 pre-orders. But what are the secrets of Tesla? Or what are the secrets of Elon Musk, the founder of Tesla?

Elon Musk decided to see if a $30 food budget could get him through a month.

He bought mostly hot dogs and oranges in bulk, and would occasionally switch it up with some pasta and canned tomato sauce. Satisfied by his success in this venture, it gave him the assurance that he didn't need a comfortable salary to survive, allowing him the freedom to pursue his lifetime goals.

Following his big dreams, Elon Musk founded Tesla with the dream of building electric cars in 2003 and has succeeded in making electric cars.

Tesla has been so successful where many other electric car companies and other cleantech companies have failed by believing in the success.

HOW TO BECOME A MILLIONAIRE BY INVENTION?

36th Story / Total 100 Stories

100 Stories of Successful Inventions

Shampoo

I remember that I used to clean my head with soap when I was a child. In the old days, only soap was needed for cleaning everything. Some people even made soap at home. Over the years, soap was followed by shampoo, then conditioner, body cleaner, and face cleaner and other cosmetics in the bathroom.

Until the mid-1930s, Westerners used mainly soap to wash their hair, but their hair was still dirty. For this reason, people used a water bottle containing coconut oil, which is easier to produce and cleaner than foam, and this was the beginning of shampoo. The sole origin of modern shampoo is still unknown, as there are many theories. The one theory we will focus on today is that it was invented in Japan after World War II.

At the time, Takeuchi Koo, who was selling wool-washing liquids, saw that children were washing their hair with soap. He then invented the first hair cleanser by removing toxins from the wool cleanser and putting spices on it. The first shampoo was intended to completely wash away the foreign matter in the hair. At that time, Japanese women put their hair oil on their hair every day to prepare their hair, and it smelled bad if they could not wash this oil completely off. As shampoos provided other functions other than oil washing, the number of Japanese women using hair oil gradually decreased.

For example, the shampoo that was invented later focused on the function of supplying moisture to the hair in addition to the original washing function. The function of the changed shampoo also greatly influenced its form.

Most shampoos were in powder form at the time, and they were mixed in warm water before use.

But in 1953, the Japanese shampoo maker Shiseido released a product called Olive Shampoo Powder, which contained natural olive oil to moisturize hair. A year later, Shiseido introduced a new shampoo in liquid form that lessened the burden of using the powder. This shampoo by Shiseido, was similar to today's shampoo, and has gained popularity among women throughout Japan.

The shampoo has become an indispensable part of our daily life, but we tend to consider the shampoo for granted.

Ads of Great King Shampoo in the Chosun Ilbo newspaper in 1934

HOW TO BECOME A MILLIONAIRE BY INVENTION?   37th Story / Total 100 Stories

100 Stories of Successful Invention   **Video Tape**

Until recently, there were many Korean video stores in Chicago. I used to borrow 20 videos every week so that I would spend the weekend watching the drama. Nowadays, people rarely borrow videos if at all and watch movies or dramas on the Internet instead.

Sony developed videotape known as Beta in 1975. Almost at the same time, JVC developed VHS videotapes. Beta video was smaller, simpler, and technically superior to VHS. Prior to the development of Beta or VHS videotape, the only film that could record analog television signals was film. Sony wanted to monopolize profits and did not share technology with other companies. JVS, on the other hand, shared technology with Matsushita (Panasonic) and the market. So it seems that better technology is not enough to win the market; you need solid cooperation with other companies to succeed in the market. Sony eventually stopped producing products related to Betamax in 2002.

Videocassettes were soon replaced by digital format DVDs, videos on the Internet, or On-Demand Video technology. Thus, even if your invention is great one, it is important to check the market to see if people really need your invention.

Beta tape is a better invention than VHS tape, but VHS tape won the market

With Video-on-demand and the Internet, the ages of Video Tapes are gone.

HOW TO BECOME A MILLIONARE BY INVENTION? 38th Story / Total 100 Stories

100 Stories of Successful Inventions

Fitbit

For many people, it can be very hard to keep dieting and exercising. It would be nice to have some personal coaches, but most people cannot afford them. Instead, Apps on smart phones or Fitbit athletic aids have become an essential exercise partner for many people.

One day when a Korean American, James Park, played video games at home, he got the idea of athletic aids from Nintendo's "Wii." In 2007, when Apple first introduced the iPhone, he co-founded 'Fitbit' and started developing wearable devices. Thanks to this, Fitbit was able to prevail in the wearable market. Apple, Nike, Philips, and other global players have stepped up to become wearable latecomers, but they have not been able to beat Fitbit's success. According to the US Fitness Tracker market share announced recently by DISTOMO (http://www.distomo.com), the popularity of Fitbit is so great that 72% of the market is occupied by Fitbit. People are wearing more Fitbit products than other good products like the Nike and FuelBand. Another successful example of Kick Starter is Fitbit. In addition to tracking steps, Fitbit will periodically email the exercise report after a user is registered with their Fitbit device. Of course, you can check the statistics at any time through the website and check it using the app on your smart phone, but it's as if the health club trainer is periodically pressuring you to exert strength on your workouts. Although it is not at the level of a social network, it is another attraction that Fitbit can give to stimulate each other with friends who exercise together. It can be a little uncomfortable to wear a Fitbit while sleeping, but analyzing the sleep pattern will help you find a way to completely refresh.

Fitbit Company, helping many people to track their exercise, is also successful financially.

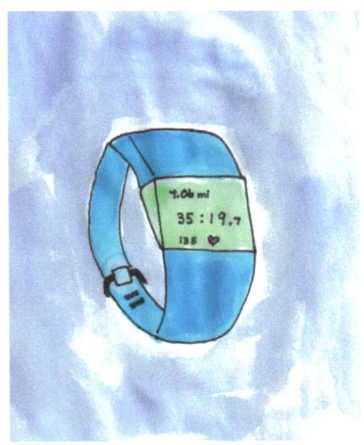

Fitbit device can help you to keep a healthy lifestyle.

HOW TO BECOME A MILLIONAIRE BY INVENTION? 39th Story / Total 100 Stories

100 Stories of Successful Inventions

Uniqlo HEATTECH

"Are there any clothing that look warm, light, comfortable and cool?"

In response to this question, Uniqlo developed Heat-generating fabric clothing: HEATTECH, lightweight and high-tech fabric actually creates heat to warm you up and keep you warm. Staying warm through a cold winter gets easier with HEATTECH. Since clothing was first created, people have gotten through the coldness of winter with either of these methods: layer clothing or thick clothing.

HEATTECH fibers absorb moisture that the body emits, and the fabric itself generates heat. Just wearing it automatically warms you—a surprising innovation in clothing technology. Moreover, the layer of air in between the fibers does not allow the warmth generated to escape, so that you can continue to feel warm as long as you are wearing it.

HEATTECH also stretches and shrinks to fit your body, allowing for unparalleled comfort and uninhibited movement.

HEATTECH is an innovative new material that UNIQLO developed jointly with global textile maker Toray.

To achieve the ultimate in warmth and comfort, Uniqlo created more than 10,000 prototypes.

HEATTECH cannot be made with ordinary equipment. The fabric is made on dedicated production lines. HEATTECH generates and retains warmth, fights bacteria to prevent odors, prevents static electricity and is extremely comfortable.

Since hitting the market in 2003, every year HEATTECH has evolved through continuous improvements that incorporate customer feedback.

Uniqlo's HEATTECH Fabric keeps you warm and comfortable.

HOW TO BECOME A MILLIONAIRE BY INVENTION? 40th Story / Total 100 Stories

100 Stories of Successful Inventions

Rubik's Cube

In dramas and movies, characters who can solve Rubik's cube puzzles are often depicted as geniuses. It is doubtful whether you really need to be a genius to solve the puzzle, but the cube, which is not easy to solve, has certainly proved to be an interesting toy.

Hungarian architect Ernő Rubik invented the Rubik's Cube, under the name Magic Cube. Then in 1980, Ideal Toy changed the name to Rubik's cube. The Cube was awarded the Best German Toy Award in 1980, as it soon became the world's best-selling toy. There are billions of Rubik's cubes sold so far and there are many kinds of Rubik's cubes being sold. (For example, a 2 × 2 × 2 pocket cube, a 4 × 4 × 4 rib cube, a 5 × 5 × 5 propeller cube, and so on).

In March 1970, a man named Larry Nichols created a 2 × 2 × 2 cube with the title "Puzzle with Pieces Rotatable in Groups" with Patent No. 3,655,201.

Rubik invented Magic Cube in 1974 and the first Rubik's Cubes were sold for the first time in a toy store in Budapest in 1977.

In September 1979, he signed an agreement with Ideal Toy to sell his Magic Cube. Later, the name was changed to "Rubik's Cube". Recently, the famous Greek inventor Panagiotis Verdes applied for a patent on how to make 5 × 5 × 5, cubes to 11 × 11 × 11 cubes. The Rubik's Cube is almost impossible to solve if you do not know the formula.

There are many reasons why the Rubik's Cube is interesting, but I can say that we can play countless kinds of games. The number of cubic arrays in the Rubik's cube is:
$(8! \times 3^8 - 1) \times (12! \times 2^{12} - 1) / 2 = 43,252,003,274,489,856,000$.

The Rubik's Cube has provided interesting toys for modern people, while at the same time it has brought a lot of wealth to developer Rubik.

Rubik's Cube invention provides some fun for many people.

HOW TO BECOME A MILLIONAIRE BY INVENTION?

100 Stories of Successful Inventions

Minecraft

Children have always loved computer and Nintendo games, and especially the game Minecraft, since 2010. Minecraft seemed addictive, but it also seemed educational in that it created a 'world of its own'. Compared to violent games that involve fighting monsters and playing "war", Minecraft was a creative game in which players collaborated to create a new world.

So who created Minecraft?

Markus "Notch" Persson, an ordinary Swedish boy who graduated only from high school, invented Minecraft. He worked as a programmer at a game company, and dreamed of making his own game someday. At the age of 12, his father divorced his mother due to a drug addiction. But Persson never gave up his dream of making his own game. Inspired by his favorite indie (independent video) game while at a game company, Persson challenged the ongoing development of games and created Minecraft in 2009. This new game received explosive responses from gamers around the world. Despite the game's low price of $26, Persson has become an instant millionaire. PayPal had blocked his account at one time because of the rapid increase in money in his account. Persson knew nothing about corporate management; he simply made the game he wanted to play, and it was announced to the world through Internet SNS. He never used marketing before. Besides, he gave the buyer of his game the fair price he thinks, not free.

By creating the best products, social media like Twitter and YouTube have created loyal customers and a solid community around the world. These together have become the driving force for Minecraft to generate sales of millions of dollars a year, and his game was eventually sold to Microsoft for $ 2.6 billion.

Minecraft Game, invented by Markus Persson, are popular around the world, was bought by Microsoft for $2.6 billion dollars.

HOW TO BECOME A MILLIONAIRE BY INVENTION?   42nd Story / Total 100 Stories

100 Stories of Successful Inventions

WhatsApp

Only a decade ago, people only needed to check the messages left on their telephone answering machines. Nowadays, there are multiple means of communication that people need to check all day, such as voicemail, e-mail, text messages, and chatting apps. These make it easy to communicate with friends and especially those who have been away for a long time.

Group chats are convenient for a group of people to contact each other at the same time. KakaoTalk, which uses 50 million people, has been used mostly among Koreans, but the WhatsApp is used more overseas and in the United States. Facebook, a leading social media company, acquired a venture company called Whatsapp in 1914 for $ 19 billion. The acquisition contract is a huge amount for venture capitals with the biggest stake in the startup, and the purchase surprised 55 employees of Whatsapp. Facebook has bought Whatsapp because the number of monthly users had already reached 450 million by 2014. Whatsapp was calculated to be much more frequently used than Twitter, which had 244,000 users at the time.

It especially absorbed the taste of young generation quickly. The largest shareholder of Whatsapp in 2014 is SEKOYER CAPITAL, and the company will invest $ 60 million (about 64.3 billion won) to earn a large sum of $ 3 billion, which is 50 times more profitable.

The story of the co-founder, American Brian Acton, impresses us with a dramatic reversal of what real success is. He worked at Yahoo and sought a way to find new jobs and partners. He also met the venture capitals, and he knocked on the door to Twitter, but both Twitter and Facebook rejected him. Then, he decided to take on the adventurous challenge of life and began his own work. The company has grown to be a giant in the industry, and Facebook, which said "No" just five years ago, gave $ 19 billion to the deal.

WhatsApp is now becoming a new icon in both online and mobile. SNS, which connects friends, connects families, and connects business customers, is a hot topic in the modern age, and it can be said that it continues to be a big hit in the future.

Whatsapp, a chatting app, was invented Brian Acton, was sold to Facebook for $19 billions.

49

HOW TO BECOME A MILLIONAIRE BY INVENTION? 43rd Story / Total 100 Stories

100 Stories of Successful Inventions

Dyson's Vacuum Cleaner

The hardest part of inviting friends and family members is cleaning the house. Usually, it is up my job as the "man of the house" to clean the stairs and the corners of the house using heavy vacuum cleaners. When I asked my friends about good vacuum cleaners, I get the same answer – buy Dyson vacuum. Why is Dyson Vacuum Cleaner so popular?

James Dyson, developer of the Dyson cleaner, had always wanted to create a better product and has proven that innovative thinking and persistent engineering can truly produce results. In 1978, Dyson discovered that the filter in the Ballbarrow's interior painting chamber was often clogged with fine paint particles, as if the dust bag in the vacuum cleaner was clogged by dust particles. To fix the problem he designed a plant cyclone tower that uses centrifugal force to filter paint particles, a force that is more than 100,000 times stronger than gravity. Dyson decided to apply the same principle after five years of research and 5,127 prototypes and to produce the first Dyson Vacuum Cleaner. In the case of Dyson's cleaner, it is said to remove fine dust using Dyson's proprietary technology called 'radial root cyclone'. The powerful centrifugal force called the 'radial root cyclone' separates fine dust particles in the air and ensures 99.9% efficiency.

For most common cleaners, the more you continue to use them, the more fine dust will be released and the suction power will become weaker. Yet Dyson's vacuum cleaner maintains suction power even after continued uses. Dyson is very effective at reducing dust and removing allergies, with features such as strong suction power certified by the Asthma Association.

Despite its higher price compared to its competition, Dyson is ranked second in market share after Hoover vacuum cleaner, with annual sales of $ 2 billion and profits of $500 million.

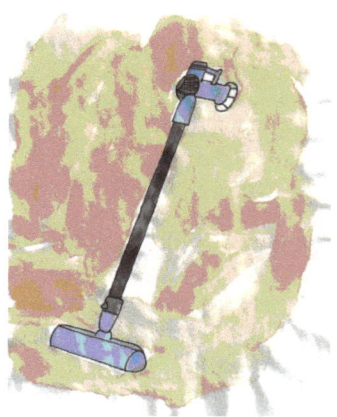

Vacuum Cleaner

HOW TO BECOME A MILLIONAIRE BY INVENTION?   44th Story / Total 100 Stories

100 Stories of Successful Inventions

Rice Cooker

Back in my childhood, my family cooked rice in a pot. I always had to remember the perfect ratio of water to rice. Nowadays, I pour the water in up to the indicated mark in the electric rice cooker, press the "START" button, and the rice cooker can start cooking perfect rice. Not to mention that it keeps the rice warm for several days. Moreover, it is very easy to cook different kinds of rice: white rice, brown rice, sticky rice, etc.   Who invented this electric cooker?

Zojirushi Company in Japan invented it in the 60s in order to solve an issue from a customer's complaint. A female worker, who had to work late at night, went back home to cook rice with an electric rice cooker. She was tired of waiting for a meal and she fell asleep. An odd smell woke her up, and she found that the rice cooker had burned the rice lightly. The employee did not think about her mistake, but called the electronics company and said, "Your rice cooker has some problems! The rice has been burned and the rice has been destroyed, so please make up for it."   The company apologized for it, worked very hard to solve the problem with the company's executives and invented a biometal (a thermostat made by attaching iron with different thermal expansion coefficients). That was when the electric rice cooker was made for safe use. If the woman had not called the electronics company or the company did not take the consumer's protest seriously and the company did not try to find the solution and the invention would have been delayed.

Thus, ironically, without the complaint of the black consumer, the company would not have seen the success of its rice cooker with warmer function for many years.

Sometimes customers who think they deserve better or who do not take reality for granted become stimulants for new inventions for the company.

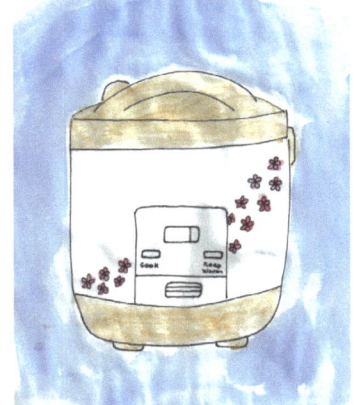

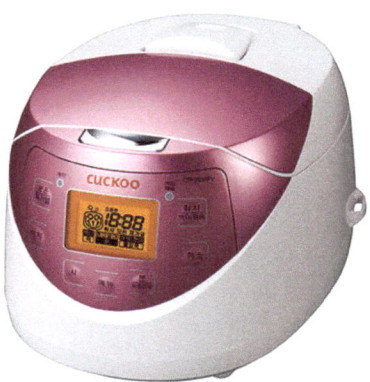

A black consumer helped the rice cooker company to develop warmer function without burning the rice.

HOW TO BECOME A MILLIONAIRE BY INVENTION?   45th Story / Total 100 Stories

100 Stories of Successful Inventions

Nintendo WII

I was first interested in Wii when I heard that it was a sports game, which required not only using my fingers to play, but also my entire body for exercise. Thus, I bought the Wii game to try it out myself.

The Wii is another Nintendo product after the Nintendo DS, and has sold over 100 million products worldwide. The success of Wii is its 'new game experience using motion sensors'. This was enough to make consumers more interested in the game, and even succeeded in attracting consumers who were not interested in the game because it was combined with the keywords 'health - fitness - exercise'. The Wii Fit package includes many sports games for bowling, golf, tennis, table tennis, boxing, dancing and so on. The program "Wii Fit" makes use of Balance Boards to improve physical fitness through various exercises such as aerobic exercise, yoga, and strength training. Naturally, it is far less of an exercise than outdoors, but it is a great development for sedentary families, including those who sit on couches and eat potato chips while watching TV.

Shigeru Miyamoto, director of Nintendo, figured that if he enjoyed the game, others would too. He elaborates that "we should not make something to sell something very popular, but to love something, and make something that we creators can love. It's the very core feeling we should have in making games."

Miyamoto wanted players of the game to feel the same satisfaction from it that the developers felt themselves.

Wii Fit helps lazy gamers to exercise.

HOW TO BECOME A MILLIONAIRE BY INVENTION?   46th Story / Total 100 Stories

100 Stories of Successful Inventions

Blue Jeans

Blue Jeans are comfortable and sturdy for everyone to enjoy. But did you know w that they were originally made of cloth that was used to make tents?

Levi Strauss invented Blue Jeans when he was at crisis. In the 1850s Gold Rush, an enormous amount of gold was poured into San Francisco in the United States. Numerous people from all over the country gathered for gold, and the city became a tent village. Strauss made money by producing tough cloths that people used to stay in these tents. One day, a merchant came and ordered an enormous amount of fabric from Strauss to deliver to the army. Strauss expanded his factory and increased the number of workers, producing cloths day and night. Suddenly, the supply route to the army was blocked and Straiss was in debt. He put a cloth that was piled up in a warehouse at a low price, but no one bought it. Disappointed at the failure, he went to a bar and observed the miners wearing ragged pants. "It would be nice if there were tougher work clothes." From that moment, the tent fabric was then reborn as work pants and Strauss called the suit "Blue Jeans". The sturdy and comfortable blue work clothes were sold to miners, and they quickly became popular with the general public.

People around the world now enjoy wearing 'Blue Jeans', an invention that came from the idea of putting an item to different use.

Blue Jeans, comfortable and sturdy, were invented using leftover tents fabrics.

HOW TO BECOME A MILLIONAIRE BY INVENTION?

100 Stories of Successful Inventions

Water bottle with Compass

When I was a child, I used to climb the mountains in the suburbs of Seoul. In the United States, I used to go the Smoky Mountains after 10 hours of driving. A water bottle and a compass are both essentials for hiking and climbing. When an invention that combined these two things together was created, it made a big hit. This water bottle with the compass, similar to the invention of pencil with the eraser, was also recorded as a global invention.

Yamashita was an expert on mountain climbing and was experienced enough to even conquer the summits of mountains in Japan.

Sometimes, Yamashita lost his way during the climbing and he did not bring a compass on the trip.

It was a big day, and Yamashita did not prepare food, but it was getting dark already. He had a cold water bottle hanging on his waist, and he thought he would not have survived without it. First, he got the water bottle and opened the bottle lid. At that moment Yamashita got a brilliant idea: imagine how helpful it would be to put a compass on the lid of a water bottle.

It was commercialized immediately after Yamashita applied for a patent, and it became a popular item among hikers and climbers.

Yamashita became an inventor with the fame and rich with it.

Mountain climber invented water bottle with compass.

HOW TO BECOME A MILLIONAIRE BY INVENTION?   48th Story / Total 100 Stories

100 Stories of Successful Inventions

Toothpaste Dispenser

When I was a child, I used to spend summer vacation at my grandmother's house. There would be a sea salt for tooth brushing in the sink in the bathroom. Then one day, a new tube of toothpaste from neighboring grocery stores appeared in the bathroom. When I used toothpaste after I used to brush my teeth with salt, I found the smell was cool and comfortable. Tubes are conveniently used to squeeze out toothpaste without thinking. Whose invention is this?

The inventor of this invention is Dr. Suwad Sheffield of Washington. Sheffield, who had to solve the discomforts of life around him, had no choice but to use the morning toothpaste. At that time, the toothpaste was not contained in the tube as it is now. In other words, it was unhygienic because the toothbrush was inserted into the jar containing the toothpaste. 'Is there any way I can use more hygienic toothpaste?' People used toothpaste in the jar without any complaints, but Dr. Sheffield was not the one. He thought he needed a way to store and use his toothpaste more hygienically. So he contacted the company that produces the toothpaste to propose more hygienic toothpaste, but the company that produces toothpaste sent me a reply that there is no other way. Dr. Sheffield has been working on hygienic toothpaste from time to time. However, it was not easy to invent new hygienic toothpaste.

Months passed, but Sheffield's research failed to produce any results. Then one day, he observed how painters squeezed the paint out of a tube to draw a picture. At that instant, Dr. Sheffield had a brilliant idea in his head. If he put toothpaste in a tube, it will be hygienic! Thus the tubular toothpaste was invented in 1892. The result was a great success. Dr. Sheffield's tubular toothpaste was popular and hygienic, and was sold out immediately.

Tubular Toothpaste Dispenser

HOW TO BECOME A MILLIONAIRE BY INVENTION?   49th Story / Total 100 Stories

100 Stories of Successful Inventions

Corning Ceramics

Ceramics are inorganic and non-metallic solids that are safe by heat and cooling activities. It is made from natural raw materials such as clay and has been used as a container. On the other hand, fine ceramics are made by using high-purity artificial raw materials and are used in various applications such as electronic materials and precision machine materials. Unlike metals, ceramics do not conduct electricity well but also resist high temperatures unlike organic materials.

If you overcome only the fragile shortcomings of ceramics, you can use ceramics for automobiles, machinery, aviation, shipbuilding, power generation, dentistry and medical care.

One day in 1952, Don Stookey, an employee of Corning Glass Works, thought he had ruined an experiment. A sample of photosensitive glass was placed in a furnace and set at a temperature of 600 degrees. When the blast furnace returned, the incomplete controller raises the temperature to 900 degrees. Stookey opened the door anticipating a melted glass drop and a ruined furnace. When he tried to remove the plate, the sample he had inserted fell off the forceps and fell to the floor. Was it broken? Though he did not realize at first, he invented the first form of artificial glass-ceramics, which he later called a reinforced heat-resistant glass, or pyroceram. It is lighter than aluminum but tougher than high carbon steel and several times more powerful than ordinary drink glass. CorningWare introduced a line of cosmic-era dishes; CorningWare that were break-resistant and CorningWare became the best seller worldwide.

In 2007, the gorilla glass, which is also break-resistant, was developed and delivered to Apple in large quantities for smart phone LCD displays and touch screen.

Don Stookey accidentally invented ceramics, which are promising for the future with the unlimited applications.

CorningWare is famous for break-resistant kitchenware.

HOW TO BECOME A MILLIONAIRE BY INVENTION?   50th Story / Total 100 Stories

100 Stories of Successful Inventions

Smart Phone

Most people check his or her smart phone frequently for email or messages. Some people cannot sleep without the smart phone at all. A few addicted people are spending more than four hours a day on the smart phones. What happened to modern people?

A smart phone is a mobile personal computer with a mobile operating system with features useful for mobile or handheld use. Smart phones, which are typically pocket-sized (as opposed to tablets, which are much larger in measurement), have the ability to place and receive voice/video calls and create and receive text messages, have personal digital assistants, an event calendar, a media player, video games, GPS navigation, digital picture camera and digital video camera. Smart phones can access the Internet through cellular frequencies or WiFi and can run a variety of third-party software components ("apps" from places like Google Play Store or Apple App Store). They typically have a color display with a graphical user interface that covers more than 76% of the front surface. The display is almost always a touchscreen and sometimes additionally a touch-enabled keyboard like the Priv/Passport BlackBerrys, which enables the user to use a virtual keyboard to type words and numbers and press onscreen icons to activate "app" features.

In 1999, the Japanese firm NTT DoCoMo released the first smart phones to achieve mass adoption within a country.

Smart phones became widespread in the late 2000s. Most of those produced from 2012 onward have high-speed mobile broadband 4G LTE, motion sensors, and mobile payment features. In the third quarter of 2012, one billion smart phones were in use worldwide. Global smart phone sales surpassed the sales figures for feature phones in early 2013.

With Steve Job's introduction of the original iPhone in 2007, Apple's goal for the first year of the iPhone was to capture 1% of the worldwide cell phone market. The company achieved that goal already and now stands at somewhere between 20% and 40% of the market. They say that Steve Jobs is an expert in marketing and promotion and an artist who know how to combine many inventions into one iPhone. It is estimated that more than 1 billion iPhones were cumulatively sold worldwide until 2017.

Smart phones have changed the lifestyles of modern people including you and me.

HOW TO BECOME A MILLIONAIRE BY INVENTION?

100 Stories of Successful Inventions

Sweetener

From time to time, I think back on my childhood and I miss the taste of the snacks my mother used to make. She used to bake potatoes and corn as a snack, and then put a little bit of sweetener in it. As well as for industrial use, the sweetener with the strong sweetness has been added to the children's snacks. How was artificial sweetener born?

In the laboratory at Johns Hopkins University in 1879, the chemist Constantin Fahlberg discovered Saccharin. "How did I discover saccharin?" He said. "Well, it was partly by accident and partly by study. I had worked a long time on the compound radicals and substitution products of coal tar, and had made a number of scientific discoveries, that are, so far as I know, of no commercial value. One evening I was so interested in my laboratory that I forgot about my supper till quite late, and then rushed off for a meal without stopping to wash my hands. I sat down, broke a piece of bread, and put it to my lips. It tasted unspeakably sweet. I did not ask why it was so, probably because I thought it was some cake or sweet meat. I rinsed my mouth with water, and dried my moustache with my napkin, when, to my surprise the napkin tasted sweeter than the bread. Then I was puzzled. I again raised my goblet, and, as fortune would have it, applied my mouth where my fingers had touched it before. The water seemed syrup. It flashed on me that I was the cause of the singular universal sweetness, and I accordingly tasted the end of my thumb, and found it surpassed any confectionery I had ever eaten. I saw the whole thing at once. I had discovered some coal tar substance which out-sugared sugar. I dropped my dinner, and ran back to the laboratory. There, in my excitement, I tasted the contents of every beaker and evaporating dish on the table. Luckily for me, none contained any corrosive or poisonous liquid. One of them contained an impure solution of saccharin. On this saccharin I worked then for weeks and months till I had determined its chemical composition, its characteristics and reactions, and the best modes of making it."

It was the birth of artificial sweeteners that can substitute sugar! It was hundred times stronger than sugar in the same amount. He did not hesitate to take his place and gave thanks to God. He succeeded in synthesizing a substance called Saccharin in the lab. Saccharin has spread all over the world at once, and was especially popular as a substitute for sugar because of its low calories.

Boiled Potato Sweetened with Sweetener

HOW TO BECOME A MILLIONAIRE BY INVENTION? 52nd Story / Total 100 Stories

100 Stories of Successful Inventions

Penicillin

Hundreds of years ago, the human life span was only 20 to 30 years old. Three out of ten children died before they were 1 year old, and about half died before the age of 10. This was due to diseases such as smallpox, measles, malaria, cholera, diarrhea, pneumonia, etc.

Louis Pasteur and Robert Koch were the ones who revealed that most of human diseases are due to microorganisms. However, the ability to repair microbial diseases and increase the longevity of humans is due to antibiotics that are not harmful to humans.

The first penicillin was discovered in 1928, in a London laboratory by Alexander Fleming who was studying grape-like germs, a common cause of swelling in children at that time. Among the 7-8 glass dishes covered with gelatin, blue mold was formed on the one of the toxic gelatin. Fleming decided that the experiment was wrong and found a stranger phenomenon when he lifted a moldy plate. The germs that had spread on the plate were gone without any reason. When the bacteria disappeared like this, something obviously had a powerful sterilizing power. Fleming decided to examine the blue fungus on the plate and soon realized that his mistake was the result of such a phenomenon. He came out of the lab with a lid on his plate, forgetting to stop observing the germs cultivated in a dish under a microscope. By the way, accidentally, the spore of the mold was grown in minutes. "Yes, perhaps this discovery may give great hope to all mankind." Fleming first cultivated the mold spores on agar on a glass plate to cultivate. Fleming began experimenting with a pumping heart. As Fleming had anticipated, all the germs except typhoid fever and E. coli were all killed by the mold. Fleming, with confidence in his research, published a paper entitled "The Nature of the Fungus Cultures on Cells, Particularly the Use of Influenza Bacterial Isolation". However, his paper was not recognized at the time, but Fleming had continued to study penicillin ever since.

In August 1942, the mass production of penicillin was finally achieved, many diseases could easily be cured, and the average life expectancy of humans began to increase. In 1945, Alexander Fleming was awarded the Nobel Prize for Medicine.

Alexander Fleming's Penicillin made us live longer.

HOW TO BECOME A MILLIONAIRE BY INVENTION?

100 Stories of Successful Inventions

Anesthesia

When I watch a war or action movie, there are often scenes where the heroes take out the bullets in the body with a knife with a blink of an eye. The heroes in the drama and the movie do not die even if they are shot, but take out the bullet as if they don't fee pain at all. Heroes are not supposed to feel the pain in the movie. In reality, if a person removes a bullet or undergoes surgery without anesthetic, they will suffer tremendous pain.

When was anesthesia invented for dental and surgical patients?

The ancient Egyptians used poppy seeds during surgery, the Chinese used cannabis to ease pain, and in the century medical books, patients were advised to drink alcohol, such as wine.

By 1730, finally, the German scientist Provenius found a hypnotic effect that kept the etheric gas in a pleasant hallucinatory state.

In 1799, British chemist and inventor Humphrey Davy discovered that nitrous oxide makes him unbearable and called this gas "laughing gas." He experimented with nitrous oxide and found that he had observed three stages: 1 painlessness, 2 delusions, 3 stage surgical anesthesia, and 4 stage respiratory tract paralysis.

In 1846, William T.G. Morton, a dentist and a surgeon, used an ether to anesthetize and successfully open the tumor.

In 1847, chloroform was used as a new anesthetic, which was used when the Queen Victoria of England gave birth to her seventh child.

After that, various anesthetics such as nitrous oxide, ether, and chloroform began to be used commonly by surgeons and dentists worldwide.

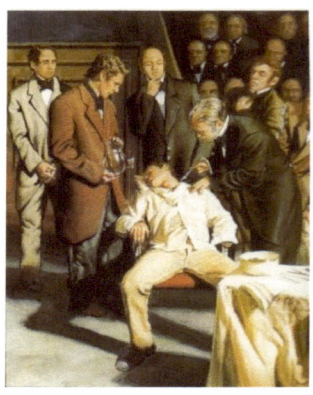

Morton used anesthetics to conduct tumor removal surgery in 1846.

HOW TO BECOME A MILLIONAIRE BY INVENTION? 54th Story / Total 100 Stories

100 Stories of Successful Inventions

Rubber Tires

What is the best invention of human history? Many say that it is the wheel. With the wheel, man could move around heavy objects. With the wheel, man could invent cars, trains, airplanes, etc.

If the wheels advanced human history, the rubber tire is an invention that made the development of the 20th century possible. Bicycles and cars were born in the mid-19th century, but all the wheels were all made out of iron or wood instead of rubber. Nowadays, rubber is used in many aspects of life such as in factories, automobiles, ships, and airplanes.

Then, in 1839, American inventor and entrepreneur Charles Goodyear invented the rubber vulcanization method, which added sulfur to increase the elasticity of rubber. Goodyear was an individual who looked at anything with the eye of invention. Especially in the case of rubber, he was crazy. He also wore hats, clothes, shoes, and gloves all made of rubber and he was regarded as a rubber madman. Goodyear, who was crazy about rubber, invented sponge rubber tires with air. It so happened that rubber was invented one day at lunchtime. "Honey! How about this bread?" The bread made by his wife was completely different from the bread he had eaten. That is, the bread was soft and the size was much bigger than the old bread. Goodyear asked, "How did you make it?" His wife answered, "I just put a baking powder." At the moment Goodyear came up with a gently inflated rubber, a sponge rubber.

'Nothing is impossible!' said Goodyear and he put the blowing agent in the rubber liquor. It was a great success. Goodyear also received the patent for the invention and all the profits with the invention.

Inflated Tire Invention was inspired by baking soda and soft bread.

HOW TO BECOME A MILLIONAIRE BY INVENTION?   55th Story / Total 100 Stories

100 Stories of Successful Inventions

Copy Machine

Laser printers are common in the office. Compared to noisy dot matrix printers and inkjet printers, laser printers print quietly, quickly and without smudged image. Moreover, prices have now dropped from a few thousand dollars to a few hundred dollars and more people can afford to buy laser printers.

Gary Starkweather, a researcher working on a high-speed fax machine at Xerox in the late 1960s, created the laser printer. "One day in 1967, I sat in the lab looking at this huge central computer. Instead of copying someone else's work, like faxing, how about using a computer to create the original? I started to think. One day I woke up in the morning and thought. How about direct printing?" He used a new technique called laser to project a large amount of light onto paper to create an image. This is how laser printing was born.

So, in 1977, Xerox and Gary Starkweather developed the 'Xerox 9700' by combining scanning technology with existing copying technology.

The world's first laser printer was a giant leap in Information Technology history with its breakthrough speed and cost-effective maintenance.

The Xerox 9700 was the first printer to instantly print characters, formatting, and various graphic images typed on a computer. It worked at a remarkable speed of 120ppm at the time of release and quickly and flexibly processes graphics and text mixed documents. It has earned a reputation as a standard for printers, with a reputation for revolutionizing document processing.

The first laser printer Xerox 9700, which has brought exceptionally fast and high-quality text and graphic printing to the business and consumer markets.

HOW TO BECOME A MILLIONAIRE BY INVENTION?   56th Story / Total 100 Stories

100 Stories of Successful Inventions

Electric Cart

An electric cart is a cart equipped with an electric motor and navigational controls. It includes a seat (often equipped with an occupant seat switch activating movement of the motorized shopping cart from the occupant's weight) thereby also making it a motorized wheelchair, and it has a rechargeable battery that can be charged by plugging in the device when not in use in order to maximize usage.

Supermarkets provide motorized shopping carts for those with permanent or temporary physical disabilities who may have difficulty walking through a large store or pushing a regular cart. Many of the customers who use motorized shopping carts are not full-time wheelchair users, but find shopping easier using the device since a regular cart may be harder to push, especially when filled with merchandise, and walking through a large store may be cumbersome for one who is able to walk only short distances on their own power.

Yogurt is a popular drink in Korea, and ladies riding electric carts with these drinks are often seen in the streets. They deliver Yogurt and other dairy products to customers everyday. Even today, the number of Yogurt sellers is estimated to be about 3,000, and they travel on their carts for an average 5km per day. In December 2014, Yogurt developed a cart for delivery to increase the convenience and to increase sales. At first, there were many skeptical views. However, the introduction of 5,000 electric carts, the sales increased exponentially. Yogurt sellers have begun to focus on consumer sales while avoiding walking around. According to an official from Korea, "Most of the Yogurt sellers who use electric carts are showing a boost in sales, though the figures vary depending on the introduction period." It was natural that if you ride a cart instead of walking, it will be a big help for the sellers carrying a heavy box of Yogurt drinks. The electric cart uses large-sized lithium-ion rechargeable battery. If you charge only once a day, you will have enough power for your daily activities.

Also, Korean Yogurt 's "Cold Brew by Barbinsky" became a popular product that sold about 100,000 a day after its launch. Average daily sales are 200 million won.

Convenient Electric Cart for Mrs. Yogurt

HOW TO BECOME A MILLIONAIRE BY INVENTION?

100 Stories of Successful Inventions

SIM Card

If you ask Google about something, the answer will be out in a few seconds. It's like Google knows us well or too much, like Big Brother of George Orwell 's novel, '1984'. They can see our movements in real time through Google accounts and phone apps. It's because everything we do is in Google's palms.

With the inventions of SIM cards and GPS (location tracking) chips, which are installed in the phones, provide many convenient functions, but in return, we expose our information to phone companies and Google. The SIM card is a small card that is smaller than a nail, but it has the power to expose our identity, finances, and communication. It's on the credit card, and it's on the phone, and in Korea, it's even on transportation tickets. You can encrypt your personal information in this tiny card and put in a small operating program. Through this program, our money goes from the bank or credit card company to the store.

Munich's smart card maker Giesecke & Devrient made the first SIM card in 1991, and the first 300 SIM cards were sold to a Finnish wireless network operator Radiolinja Corporation.

SIM cards have their own number, a fixed number ICCID (19 digits) and IMSI (15 digits) that varies from one subscriber to another. If you want to upgrade your phone to another one, you can just plug a SIM card from old one to new one. Without a SIM card, you cannot use most smart phones or banking services, such as calls and text messages, but you can call emergency numbers. If you call an emergency number with a SIM card-free terminal, you will not see the phone number in the Emergency Call Status Room, and the phone's unique number (IMEI) is displayed instead to identify the caller. It was originally used only in Europe, but it became mandatory in the United States.

We are using multiple SIM cards at the same time, including those on cell phones and credit cards. In other words, it has become pretty easy to expose our information to the Internet.

Your smart phone with SIM and GPS chips can be your best friend and your worst privacy enemy.

HOW TO BECOME A MILLIONAIRE BY INVENTION?

100 Stories of Successful Inventions

Rabies Vaccine

Dogs are cute and friendly, and they are often ranked first out of preferred companion animals. However, there was a saying, "There is no medicine if you are bitten by a mad dog," As this saying suggests, rabies was frequently a subject of terror. With this disease many people bitten by mad dogs died. Symptoms include an abnormal sensation in the area of the bite, sensory paralysis, and poor wound healing. The patient feels hypothermia, headaches, vomiting, and is easily made anxious, causing muscular rigidity and convulsions. Rabies can also cause diarrhea, induce comas, and end up being fatal if the patient is paralyzed during breathing. The incubation period is about 3 to 6 weeks, and most patients will develop symptoms and die about 4-10 days.

From the mid-nineteenth century, scientists called Louis Pasteur, the father of microbiology, began studying rabies. Pasteur heated the anthrax bacterium, a pathogenic bacterium of rabies, to weaken the disease-causing force. After that he experimented with animals and the results were successful. Soon after, Pasteur had the opportunity to experiment with the results of the research. It was a teenager and a mother who had been bitten by a madman. The boy's condition was very poor, but Pasteur began to worry as he had to conduct his first experiment with a person. But his fellow doctors actively supported him, and the result was a great success. The boy, who was injected once a day for 14 days, no longer had symptoms of rabies. This was in July 1885. Since then, Pasteur's rabies vaccine has succeeded in treating people, and it has become increasingly known to the world. Now thanks to Pasteur, "If you are bitten by a mad dog, you can go to the hospital and get treatment."

Rabies vaccine is given to people with increased risk of rabies.

HOW TO BECOME A MILLIONAIRE BY INVENTION?   59th Story / Total 100 Stories

100 Stories of Successful Inventions

Ramen Soup

I used to eat ramen soup after swimming, skating or picnic at the park because hot and spicy ramen soup makes my body warm.

Ramen is a noodle soup with popularity has grown so much that it is now eaten all around the world.

Who invented such a delicious food?

A businessman named Ando Momofuku in Japan, a paradise of ramen soup, developed Ramen. At this time, 'Ando' thought that he could use wheat flour to develop a dish that was as good as rice. It was not as easy as he thought. The two-month study was a failure. After years of research failures, he was depressed and desperate. He lived a day to day with his tough reality.

One day Ando ordered liquor and tempura in the restaurant and the owner of the restaurant was splashing tempura. Ando kept looking at the master frying tempura. It was the moment when the moisture in the wheat flour came out instantly, and the small number of small holes occurred in the food when the fried food was cooked. Then the idea or flash of genius came to Ando's mind, "Yes, this is the principle of cooking the tempura." Ando hurried to the lab and started experimenting. First, he made flour into noodles and fried them in oil. Then the water in the noodles evaporated and a small hole was formed in the inside and was dried. Then, he poured hot water. This time, hot water was entered into a small hole and it became a delicious noodle.

He was exited, but he kept it for a few days before commercializing ramen. It was the moment when ramen was finally born and the life of Ando was turned around and Ando became the legend.

Ramen Noodle Soup with Kimchi

Silver statue of Momofuku Ando in Ramen Museum in Japan

HOW TO BECOME A MILLIONAIRE BY INVENTION?   60th Story / Total 100 Stories

100 Stories of Successful Inventions

Microwave Oven

The microwave oven, which is essential for warming up leftover pizza from the day before, can be easily found in many kitchens. With the microwave oven, we can cook convenient one-minute lunch or three-minute gourmet dinner packs from the stores.

Microwave ovens are based on microwaves, electromagnetic waves with electric field changes from 1 billion to 30 billion times per second. This causes water molecules in food to move quickly. Because of this movement, there is friction between the molecules, the temperature rises and the food heats at that temperature. This has the advantage of making the inside and the inside of the food even better, and the cooking time is greatly reduced.

The inventor of the microwave oven is Percy L Spencer (1894-1970). He could not graduate from elementary school because of the difficulty of His family. After working at the ironwork factory, he moved to the 'Rayton Company' where he built the tube. Although he did not receive a proper education, he had an excellent brain and became an inventor. The invention of the microwave oven is a result of chance adding to his sincere efforts.

One day in 1945, Spencer, who was devoted to research by the tube, was surprised when he put his hand in his pocket. The chocolate in the bag that he had put in the street was not hot, but it was all melted away. The same day he put chocolate in my pocket the same day. Spencer, who believed that there was 'obviously some reason,' began his research.

After long deliberation, he noticed that chocolate melted because of microwaves blowing from a vacuum tube. Using this principle, Spencer created a microwave oven that cooks food without burning and received a patent. The first microwave oven (shown below) with 150cm height and 340kg weight was mainly used in large restaurants, trains, and ships, rather than homes. With the development of technology, microwave ovens have become smaller. It become more diverse and functional in a way those children who do not usually cook can easily heat up the prepared or leftover food.

Microwave Oven is an invention that even children can prepare food with.

HOW TO BECOME A MILLIONAIRE BY INVENTION?

61st Story / Total 100 Stories

100 Stories of Successful Inventions

LED Lights

When I was young, I accidentally touched a hot incandescent light bulb and I still remember the pain from the burn. Now, the incandescent light bulb is disappearing in the shops and around us and replaced by LED light bulbs. LEDs are cold with no risk of burning, use less electricity, and posses a longer life. How was it invented? The history of LEDs (light emitting diodes), which are being used in our daily lives, such as remote controls in the home, mobile phones, TVs, electric sign boards, and light bulbs, goes back about 100 years.

In 1907, a British radio engineer, Henry Joseph Round, investigated the electrical properties of a metal-semiconductor silicon carbide (SiC) rectifier as a substitute for a vacuum diode, and accidentally discovered the emission of light. The LED is basically a monochromatic light source. Depending on the combination of compounds, different light is emitted from the long wavelength region to the short wavelength region of visible light. The LED has been mainly used as an information transmission display such as a traffic light and an electric signboard. Since the development of a blue LED in the 1990's, it has finally become possible to implement a white LED. The reason why blue LED is important for realizing white LED is closely related to the characteristics of light. As we learned in our school days when we were young, when we mixed all the three primary colors of light (red, green, and blue), we get a magical white light. Because LED is a semiconductor, it is possible to control light color, color temperature, brightness, etc. Incandescent lamps, one of the best inventions in human history, and fluorescent lights, one of the most widely used lights in modern society are disappearing because of the problems of high power consumption, mercury use, and environmental pollution and short life span. Because of this, it now announces the ban on the production, import and sale of all kinds of incandescent lamps worldwide. Instead, they are replacing LEDs with high-energy efficiency and eco-friendly features.

Now our generation has become the last generation to know incandescent lamps and our future generations would ask about primitive lighting of "incandescent lamps", which was too hot and consumes too much energy.

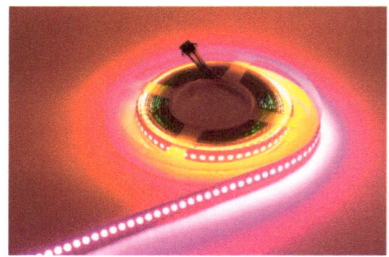

LED light strip

HOW TO BECOME A MILLIONAIRE BY INVENTION?   62nd Story / Total 100 Stories

100 Stories of Successful Inventions

Amazon's 1-Click Shopping

In the United States, it is so easy to online shop because of Amazon OneClick and PayPal. This is because users are required to input a minimum amount of information and apply various security technologies at the back end. This is possible because the security regulations for card companies in the US are different from those in other countries.

OneClick was also applied to Apple iTunes in 2000 as Amazon patented the Internet shopping process registered with the United States Patent and Trademark Office. The method is simple. When a user clicks on "Buy now with 1-Click" via Amazon.com and its affiliate site, the billing information and address are entered and payment is made on the designated credit card so that the goods can be delivered immediately. Amazon uses 'cookies' that contain information about the files sent and received via a web browser so that the next payment can be made without the same input. PayPal acts as a kind of virtual account, which means you can access the PayPal website and sign up, enter your credit card information, and use it once you are authenticated using an approved text message. After that, you can pay by simply entering your ID and password after clicking PayPal payment at the shopping mall. The two payment methods, which seem fairly unstable in other countries, can be widely used in the United States because of the authentication of the Data Security Standard (DSS) that the Payment Card Industry (PCI) proposes to safeguard cardholder information. Amazon OneClick, PayPal, and others are regularly audited for security through PCI-DSS certification. PayPal also has a bug bounty (a bounty payment if it finds defects) that pays out a reward if it finds out security vulnerability from its services. A remote code injection attack has a bounty of $ 10,000, SQL injection of $ 5,000, and authentication bypass of $ 3,000. Amazon and PayPal's policy of ensuring that goods are easily bought first and then secured thoroughly afterwards is the prudent attitude of the US financial supervisory authority and Amazon's good marketing strategy.

Amazon's 1-Click Shopping is a good invention and good marketing strategy.

HOW TO BECOME A MILLIONAIRE BY INVENTION?   63rd Story / Total 100 Stories

100 Stories of Successful Inventions

3D Printer

Nowadays, you can watch movies in 3D in movie theaters. If you use only special glasses from the movie theater, you will be amazed at the illusion and reflexes when the arrows or stones flying over to you.

Recently, the 3D printer has been getting popular. It is a machine that prints data existing on text or image on paper, and then prints a 3D design as a three-dimensional object. 3D printer 30 years ago, in the early '80s, the US company' 3D Systems, Inc. developed the world's first printer to print 3D liquid with plastic liquid using PE (polyethylene).

In 1988, Scott Crush, founder of Stratasys, was making toys for his daughter in the kitchen. He stuffed a mixture of polyethylene and wax into a glue gun and poured it one by one into the shape. This became the beginning of the 3D printer method, and it started the business with the patent. At present, there are a variety of materials used to create stereoscopic shapes (more than 30 3D printer ink materials developed to date), but the first ink in 3D printers was polyethylene. 3D printer ink is made up of petroleum compounds such as plastics, PE, PVC, PLA, ABS and so on.   3D printers print 3D design drawings as real objects with two main principles of 'rapid prototyping method' and 'computer numerically controlled engraving method'. 'Rapid prototyping' is a much more common way to build up a very thin layers. At this point, the thinner the layer, the more elaborate the object.

Overlapping layers of printers contain powder, liquid, and solid ink. The printed matter output to the 3D printer is immersed in a hardening agent, followed by drying and finalization.

The original purpose of 3D printers was to create prototypes. Once the product has been designed and designed, it will be sampled in advance at the factory before the product is taken. When you make prototypes like this, it is easy to know what kind of problem you have when it is made as a real product. So, 3D printers have started to be widely used in the manufacturing industry.

Now, the sophistication and completeness of 3D printer print results is advancing. It was developed not only as a sample output for prototypes, but also as much as it could be used in automobile and various industrial fields and real life.

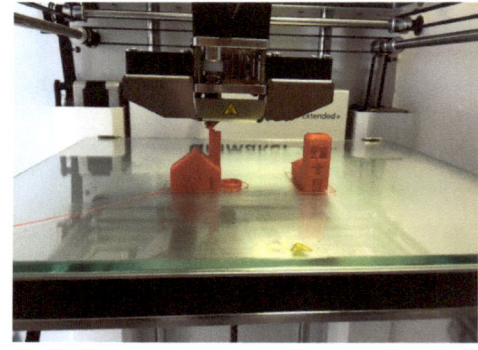

3D Printer makes the rapid prototype possible and saves time and money.

HOW TO BECOME A MILLIONAIRE BY INVENTION?   64th Story / Total 100 Stories

100 Stories of Successful Inventions

Mickey Mouse by Disney

The sixth in the world and the first Disneyland in China, Shanghai Disneyland, opened in 2016. Mickey Mouse's popularity in China was booming, with 600,000 people visiting during the trial period of just over a month.

Disney Company was founded by a poor couple who used to live quite humbly for over 90 years ago. "Honey, how do we make a living from now on?" asked Mrs. Disney. The poor Disney couple was kicked out of the room, sitting in the park and worried about their future. Walt Disney was born in a poor family and had a learning disability that he could not read until he was nine. He wanted to paint but he did not have any money, so he began to draw pictures with all kinds of bad things such as delivery of newspapers early in the morning. The advertising company was kicked out because it was not good at painting. In the end, as a homeless couple, they used to shed a tear while praying in the church and they lived in a shabby church warehouse. Disney did not know anyone but he painted hard and he was always dreaming and he did not give up. He continued to study and study painting. One day, a cute little mouse in a shabby warehouse was joking in front of them. As usual, the Disney couple used to forget about the reality of poverty by watching the cute mice that became like pets. Suddenly, the idea of drawing the mouse came to Disney, who was looking at the mice. "Yes, I have to draw the shape of the mouse in a cartoon. If this comic book comes out, it could help alleviate the sorrow of the poor. That's how Mickey Mouse was born in a world of animation. Disney rushed to publish a cartoon called Mickey Mouse. Mickey Mouse became popular worldwide, and Walt Disney became a billionaire and was able to build Disneyland. Disney says. "If it is possible to dream, it is possible to fulfill that dream, remember that all this began with a small mouse, all our dreams will come true, if there is courage to believe it."

Mickey Mouse character, created by Disney, said to be inspired by the real mice at the warehouse.

HOW TO BECOME A MILLIONAIRE BY INVENTION? 65th Story / Total 100 Stories

100 Stories of Successful Inventions

GoPro Camera

Nowadays, the smart phone has a front and back camera, so people can take pictures of others and of themselves as well. However, there are a few shortcomings to taking self-portraits while moving. There are some smart phones that are not waterproof, so it is not safe at the waterfront. Also, if someone tries to take a short video, the memory of the phone may be insufficient. There is a person who thought about this and made a big hit.

Nick Woodman, founder of GoPro. GoPro is a US start-up company that produces action cameras that are used to capture exciting moments while enjoying extreme sports such as mountain biking, snowboarding, sky surfing, water skiing, and more. In 2014, GoPro made an IPO at $ 24 a share and went up to $ 40 on 30 trading days. It was that moment when founder Nick Wood became a billionaire. Wood studied Visual Arts at UC San Diego, USA. After graduation, he established a marketing company called Funbug, but it was only met with failure, so he went on a world tour in 2002. In Australia and Indonesia, the 35mm camera was fixed in the palm of his hand with an elastic band and he took his own surf. "Let's create a camera that can capture dynamic images easily while enjoying a sports game!" GoPro started this way. Wood borrowed $ 35,000 from her mother to start a new business. His father, a founder of investment bank, raised $ 200,000 by collecting money from friends and relatives. This $ 200,000 became $ 280 million. Originally, he was trying to develop a belt that fastened the camera to his body, but he then developed a 35mm camera that was attached to his wrist. The GoPro is now a compact digital camera that can be WiFi enabled, remotely controllable, waterproof, micro SD card, dual action for general action sports, and a price tag of $ 200 to $ 400. Two years after its founding, GoPro sales have doubled every year. In 2012, Samsung sold 2.3 million units in December that year and Foxconn, an Apple iPhone maker, invested $ 200 million to take 8.88 percent of the stake. As expected, founder Nick Woodman and his wife, Gilles Scully, became very rich. Of course, Woodman's name also appeared in Forbes World's Richest People List.

GoPro Camera is popular among both young and professional people.

HOW TO BECOME A MILLIONAIRE BY INVENTION?   66th Story / Total 100 Stories

100 Stories of Successful Inventions

Overwatch Game

Some people still recognizes the computer game as an object of poisoning and regulation without accepting it as part of the cultural industry. Computer game is popular because it makes people happy.

Overwatch is a team-based First Person Shooter (FPS) game developed and distributed by Blizzard Entertainment. The game was released on May 24, 2016. There are currently 21 playable heroes, and each hero can be classified as attack, assault, defense, and support depending on their role. There are a total of 12 maps, which can be divided into occupation, escort, contest, and occupation / escort depending on the type of play. Overwatch is easy to operate once. As a result, it is easy for anyone, both young and old, to enjoy it, thus securing a large number of female users who are relatively away from shooting games. In addition, character design is diverse. Looking at the existing FPS, it is a way to shoot the enemy in search of the enemy mainly in the existing weapons of military uniform. Overwatch, however, has 21 characters with distinctive characters to show off their different outfits and weapons. This also attracted female users who liked cute things. The Korean character 'Diva' also appears. Diva has a lot of male fans with cute looks. The game was sold more than 10 million copies worldwide in one month. Tesla founder Allan Musk, CEO of the electric car, also mentioned how popular it is that even if you want Twitter to have a fast first-person shooter (FPS), he recommended Overwatch. The success of Overwatch has shown many of the common preconceived notions in many parts of the computer game market are wrong: online games and First Person Shooter games are hard to succeed and FPS, which is not a military, has few users. People should not see the game industry as a "target of addiction and regulation," but rather as a sector of the software industry. As long as the computer game give many people the joy of playing games and provide jobs for programmers, the computer game should be accepted as one area in future industry.

Diva character in Overwatch game attracted many female users.

73

HOW TO BECOME A MILLIONAIRE BY INVENTION? 67th Story / Total 100 Stories

100 Stories of Successful Inventions

Razor Shaver

When I was very young, I used to watch my father shave his face with the primitive razor blade and some soap bubble. But he used to get cuts a lot, and he often used bandages or egg white as remedies

Over the years, the performance of the razor has improved so much that men no longer have to go through the pain of skin cuts. However, even a hundred years ago, it was not easy shave without getting some skin cuts.

The inventor who brought the revolution of the razor was a regular salesman named "Gillette". "Gillette" also had a problem in shaving and he swore that "I devote myself to research for a year with the belief that I would make a new razor that would not cut my flesh." But time went by without performance. Then one day, Gillette came to the barbershop by chance. At the barbershop, Gillette discovered an important fact. It was found that the barber cut only the hair that had sprung out through the comb tooth. "Ah!! That's it! Soon we could make a safety razor!!!" He was soon using that principle to create a safe razor. "I experimented with myself and I was able to make a safe shave razor." He applied for a patent, set up a factory, and started mass production. Even a full month of full-fledged operation, there was a flood of huge orders and two years later it was exported to more than 20 countries. After 10 years, it exported to 50 countries and Gillette expanded the factory. Now the name "Gillette" is being used as a synonym for safe razor shaver.

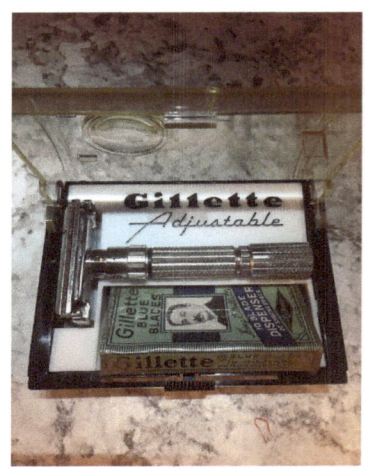

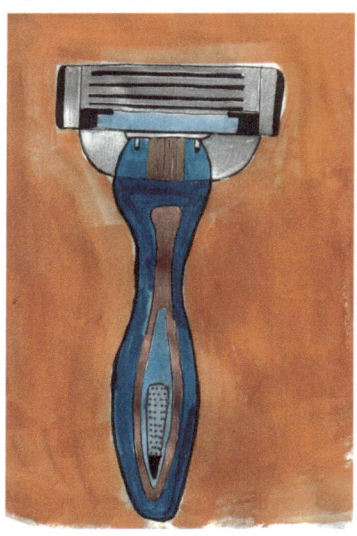

Gillette invented Razor Shaver inspired by the process of getting a haircut at a salon.

HOW TO BECOME A MILLIONAIRE BY INVENTION?   68th Story / Total 100 Stories

100 Stories of Successful Inventions

Shaved Ice (Bingsoo)

The summer in South Korea is very hot and people flock to cafés or restaurants for Bingsu. Bingsu ("shaved ice with read beans") is a popular Korean shaved ice dessert with sweet toppings that may include chopped fruit, condensed milk, fruit syrup, and red beans. The food originally began as ice shavings with red bean paste. Many varieties of Bingsu exist in contemporary culture.

The early forms of Bingsu consisted of shaved ice and two or three ingredients, typically red bean paste, rice cake, and groundnut powder. The earliest forms of Bingsu existed during the Joseon Dynasty (Former Korea in 1392–1910). Government records show officials shared crushed ice topped with various fruits.

The modern forms of Bingsu are reputed to have originated during the period of Korea under Japanese rule (1910–1945) with the introduction of a cold dish featuring red bean paste. The combination of red bean paste and shaved ice is a Korean invention.[During the Korean War (1950–1953), foreign influence led to the inclusion of ingredients such as fruit cocktail, ice cream, fruits, nuts, cereal, syrups, and whipped cream. In the 1970s and 1980s, popular ingredients included fruit cocktail, whipped cream, and maraschino cherries.

There are a variety of Bingsu types and flavors. Many bingsus do not necessarily follow tradition, and some do not include the red bean paste. Some popular flavors are: green tea, coffee, and yogurt.

Bingsoo can be found at most fast food restaurants, cafes, and bakeries in South Korea. Bingsu is also a very popular dessert at cafes in the Korea towns of Vancouver, New York City, Los Angeles, and Atlanta.

Shaved Ice is a very delicious dessert with the ingredients of fruit, milk, syrup, red beans and shaved ice.

HOW TO BECOME A MILLIONAIRE BY INVENTION?   69th Story / Total 100 Stories

100 Stories of Successful Inventions   **Nintendo Pokemon Go**

One Saturday night, in front of the headquarters of Stanford University in California, USA, 'Main Quad', where the dimly lit light was shining, a group of people were pushing their fingers on the screen of their smart phones. They came to catch Pokemon with their smart phones on the augmented reality (AR) game app called 'Pokemon Go'. Many of these young people seemed to be students, but there were occasional elderly couples and middle-aged. It is not usually a night at night because there are no libraries or laboratories, but as the storm of Pokemon is sweeping across the United States, including Silicon Valley, the people who try to get items and get Pokemon at midnight on the weekend are constantly on their feet.

Meanwhile, in Korea, gamers from all over the world are running to Sokcho to play Pokemon Go game. Sokcho has reopened its free Wi-Fi zone map that had already been created in the past.

Niantic, a US start-up launched in Google in 2015, developed the augmented reality (AR) game with Nintendo's subsidiary Pokemon Company Group. 'Pokémon Go' was released in the US, Australia and New Zealand on June 6, 2016. Nintendo's share price soared 53 percent in the last three trading days on the Tokyo Stock Exchange, bringing the total market capitalization to US $ 9 billion. Pokemon go utilizes various characters in the story of the 'Pokemon' animation, and can be played on smart phones by combining GPS and augmented reality technology. After the user launches the app on their smart phone, they can hunt out the Pokémon by traveling around. If you have nearby Pokemon or game items, you will feel the vibration on your smart phone.

If you find Pokemon on the screen, you can catch it by throwing a pocketball. Depending on the nature of the Pokémon, water-related Pokemon was supposed to be found in the surrounding rivers and oceans, and the gamers were able to find items for pocketball and to find Pokemon at places such as museums, art galleries, monuments and monuments.

Pokémon Go is a free-to-play, location-based augmented reality game.

HOW TO BECOME A MILLIONAIRE BY INVENTION? 70th Story / Total 100 Stories

100 Stories of Successful Inventions

Two-way Electric Socket

When I first graduated from university and joined Panasonic Company (Matsushita) in Chicago, I received a copy of Matsushita's autobiography book as a gift for joining the company. At the beginning of Matsushita's book, an electric socket story came out.

There used to be only single electric socket in Japanese households before the dual electric socket was invented. The dual electric socket was the foundation of the Matsushita Group, Japan's largest consumer electronics company. The hero who founded the world's greatest group with this small invention was a Japanese named "Matsushita Konosuke". In the fourth grade of elementary school, when his father failed his business career, he stopped studying, left the countryside, and started working as a clerk in Osaka City. Young Matsushita worked for 10 years in bicycle and light bulb companies and got his own electric appliance store. It was this little electrical goods store that changed his fate. Matsushita sold electric appliances, such as electric wires, plugs, and sockets in his small shop.

One day, Matsushita went out for repair job and by chance he heard that the sisters were arguing. One socket was available and the sisters wanted to use it first. Matsushita came up with a lightning idea passing through his head at that moment. To make sisters use the sockets together without fighting each other, he invented duel electric socket. Matsushita immediately applied for a patent and started making double electric sockets.

The dual electric sockets were selling like hot cakes in all the electric appliance dealers. Matsushita's little store was changed to "Matsushita Electric Company" within a year. After that, Matsushita continued inventing. Anything in Matsushita's hand would be renewed and a new patent was born. It became the Matsushita group, and later Panasonic company of the United States.

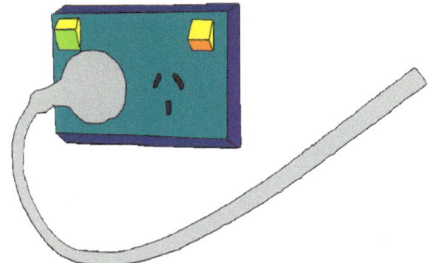

Two people can share a two-way electric socket, invented by Matsushita Konosuke, the founder of Matsushita Group and Panasonic.

HOW TO BECOME A MILLIONAIRE BY INVENTION? 71st Story / Total 100 Stories

100 Stories of Successful Inventions

Cancer Diagnostic Kit

Many people die in just a few months after getting diagnosed with pancreatic cancer. 85% of patients with pancreatic cancer, known as one of the most terrifying and deadliest cancers, die early due to its often late diagnosis and higher average mortality rate than other cancer types. Most of the diagnostic kits used to diagnose pancreatic cancer are expensive, about $ 800, and utilize outdated technology from over 60 years ago.

15-year-old American high school student Jack Andraka invented a paper sensor that could detect pancreatic cancer, ovarian cancer, and lung cancer at a cost of 3 cents a pound within five minutes of testing. Andraka said, "We decided to develop a diagnostic kit directly because we thought it was inexpensive, simple, fast and selective." He became interested in the diagnosis of pancreatic cancer when his favorite uncle suddenly died of pancreatic cancer. Andraka first explored more than 8,000 proteins over three months to find mesothelin, a protein that responds to pancreatic cancer, ovarian cancer, and lung cancer. He combined the antibody with carbon nanotubes to create a diagnostic paper. However, he could not commercialize the samples he made alone, so he applied for registration to many institutes, but he received 199 refusal letters from an existing research institutes. Finally, Dr. Johns Hopkins called him, and he was able to complete the paper sensor development. With the diagnostic kit, the pancreatic cancer survival rate could be increased from 5.5% to almost 100%. By simply changing the antibody on the sensor, people can diagnose a number of other diseases as well. Andraka, who is now an adult, are still working hard to reach his dreams of inexpensive and quick cancer diagnostic kit.

Just when the Internet is concerned, Andraka who utilized the Internet as a source of information rather than entertainment, says, "Anything on the Internet is possible. You do not have to be a professor with multiple degrees to share your theories and your thoughts to be important. It is a neutral space, and it does not matter what it looks like, age or sex. Your idea is important. It is more important for me to see the Internet as another point of view and to be able to do more there. Not just a place to share photos taken in strange poses. You can change the world!"

Jack Andraka invented a paper sensor that could detect some cancers.

HOW TO BECOME A MILLIONAIRE BY INVENTION?

100 Stories of Successful Inventions

Shake Shack Burger

If there is "In and Out" hamburger house in western California, they say that there is "Shake Shack" hamburger in eastern New York.

In 2004, New York City began taking bids to operate a new kiosk-style restaurant within the park; Meyer outlined his idea for the space, and opened the first Shake Shack in July 2004. Randy Garutti established a hot dog cart and the cart became extremely successful, and remained in operation for nearly three years. From its beginning the restaurant was not designed to be a chain, intended to be a single shop location designed specifically for New York City. However, as the original location's sales continued to grow, the group realized that there was a market for expansion.

Since its opening, Shake Shack has grown and its average store revenue of US $4 million is more than twice that of McDonald's average store revenue within the United States. Its popularity is such that in the summer at its original location, the wait in line for service can stretch to over an hour, especially on weekends when the weather is pleasant.

Shake Shack's approach in particular—sourcing high-quality natural ingredients, cooking food to order, and placing a major emphasis on the happiness of its customers and employees—both reflects and is driving real change in the marketplace.

Also, Shake Shack's eponymous products are its milk shakes, which have been reviewed as some of the best in the industry. As it has grown, the company has also added wine and bottled beers to its beverage menu. In each new location the beverage menu is customized to the local flavors of the city in which it operates.

A hot dog cart has become a global hamburger brand by re-inventing hamburgers.

HOW TO BECOME A MILLIONAIRE BY INVENTION?   73rd Story / Total 100 Stories

100 Stories of Successful Inventions

Uber

My acquaintance with a full time job drives an Uber-taxi to make up for nearly $ 1,000 of insufficient monthly income. Anyway, it seems that my friend is making good use of the idling car.

Why do people love Uber?

Because the driver has some free time for the ride and the user can ride for cheaper price.

Uber Technologies Inc. is an American technology company headquartered in San Francisco, California, United States, operating in 633 cities worldwide.

It develops, markets and operates the Uber car transportation and food delivery mobile apps. Uber drivers use their own cars, although drivers can rent a car to drive with Uber.

The name "Uber" is a reference to the common (and somewhat slangy) word "uber", meaning "topmost" or "super", and having its origins in the German word über, meaning "above".

Uber has been a pioneer in the sharing economy and the changes in industries as a result of the sharing economy have been referred to as "Uberification" or "Uberisation".

Airbnb is using the similar concept of Uber. Airbnb let people to lease extra rooms or houses. Airbnb is an online marketplace and hospitality service, enabling people to lease or rent short-term lodging including vacation rentals, apartment rentals, homestays, hostel beds, or hotel rooms. The company does not own any lodging; it is merely a broker and receives percentage service fees (commissions) from both guests and hosts in conjunction with every booking. It has over 3,000,000 lodging listings in 65,000 cities and 191 countries, and the host sets the cost of lodging.

Many Companies Are Successful in Using the Sharing Economy Concept.

HOW TO BECOME A MILLIONAIRE BY INVENTION?   74th Story / Total 100 Stories

100 Stories of Successful Inventions

Online Education

It is a great pain to go to school for many young people who love to sleep in. Especially in college there are many students who take morning lectures and decide to skip class and borrow their friend 's notes instead. But this is a waste of money and time due to expensive college tuition fees. Since the 1980s, the development of computers, the emergence of the Internet and the Internet shopping malls have been continuously growing. Then came the idea of online or Cyber University that offers lectures on the Internet as a way to earn a degree. It is a higher education system that grants a bachelor's degree or a bachelor's degree if the learner learns the education service provided by the teacher over the Internet without restriction of the time and space, and finishes certain credits. The student can take the course at a convenient time with a computer and the Internet. It has the advantage of being able to take courses at home and receive a degree without having to physically travel.

Dr. John Sperling, a professor of economics at the University of Cambridge, is a prominent example of success as an entrepreneur. In the early 1980s, about 30 years ago, he argued at the World Future Conference that "future university education will evolve into cyber universities or remote universities," but few people believed his arguments. Dr. Sperling founded the University of Phoenix, the online university was soon transformed into a super-sized cyber university with more than 100 college courses, with more than 200,000 registered students and 17,000 professors in 200 education centers.

In addition, MIT University developed a course based on basic theory as cyber content, and as a result of promoting free course service through the Internet. As a result, many people around the world, including people of all ages, genders, and races, have been able to receive these services, and this change in university education has begun to affect other top universities in the United States, including Harvard University. The commonalities of Phoenix University and MIT University are a good example of the fact that higher education is no exception in the era of the connected world. With Online education, students can save money and save time while getting education and degrees.

Online Education saves time and money and makes classrooms disappear

HOW TO BECOME A MILLIONAIRE BY INVENTION?   75th Story / Total 100 Stories

100 Stories of Successful Inventions

Metro Subway

When I am going to downtown Chicago on Highway 94, I try to avoid the roads around 8:00 AM and 5:00 pm traffic hours if possible. The Chicago highway is likely to take more than an hour because of heavy traffic. On a busy highway on the commute, the author used to think that flying cars should be invented as soon as possible.

Chicago, which has a population of three million people, is in a serious state of traffic as people commute from work to home.

The over population problem in metro areas are serious in other countries, too. The population of China is 1.4 billion. The registered number of passenger cars in Beijing, which has a resident population of 20 million, is 5.6 million. Cars are growing, but the roads are already saturated. To solve this problem, Beijing city plans to increase the proportion of public transportation such as buses and subways to 75% by 2020.

Also, Seoul metropolitan area, in South Korea, has the population over 10 million people. The Seoul Metropolitan Subway is a metropolitan railway system consisting of 21 rapid transit, light metro, commuter rail and people mover lines located in northwest South Korea.

The system serves most of the Seoul Metropolitan Area including the Incheon metropolis and satellite cities in Yogi province.

Some regional lines in the network stretch out to rural areas in northern Chungnam province and western Gangwon province that lie over 100 km away from the capital as well as Suwon.

The annual ridership of Seoul Metropolitan Subway is over 1 billion per year.

By providing transportation under the road, metro subway system solves the problem of over population in the metro areas around the world.

London Metropolitan Subway in 1861

Seoul Subway has more than 1 billion annual ridership.

HOW TO BECOME A MILLIONAIRE BY INVENTION? 76th Story / Total 100 Stories

100 Stories of Successful Inventions

Air Conditioner

Most people are familiar with the unpleasantness of humid, hot and long summer nights. Thankfully, people can be saved from this discomfort due to the invention of the air conditioner.
So who invented this device that pours cool air into the heat of summer?

This invention is the work of Willis Haviland Carrier (1876-1950). The invention of air conditioner was not intentional. Carrier started working for a company called Buffalo Forge Co. Carrier's first assignment was to study the device for drying wood and coffee. Carrier invented the first air conditioning system at the Buffalo Ironworks to control temperature and humid air circulation on July 13, 1902. This success contributed significantly to the cost savings of the company, He became the development team leader. When the drying device succeeded, he finally received his first customer request. It was a print shop close to the sea in Brooklyn, New York. The print shop, which is vulnerable to moisture, has become a mess because the prints have been scattered and messed up at the time. At the end of the study, Carrier successfully solved this problem by developing a system that can control the humidity in the air. In 1906, another opportunity came to him. This time, the problem was not 'humidity' but 'heat'. In a South Carolina fabric factory, the heat generated with thousands of additional fabrics resulted in machine failure. At the end of the study, a cooling system was developed and patented with the Apparatus for Treating Air.

In 1915, six young men who had the same intention with Carriers started Carrier Engineering Corp. Carrier remains a big and famous air conditioner company. The carrier's air conditioner, which was used only for the condition of the product at the factory, has become modified to be more comfortable air for people in their homes. The first air-conditioned building was said to be three theaters called Carrier in Texas. At that time, people were often seen in crowded theaters or in confined areas where people were overheated, vomited, or even fainted. The air conditioner of Carrier was installed in such a place, and the effect was, of course, a great success.

By 1965, more than 3 million households in the United States had air conditioners, and Carrier air conditioners are now being used in various places, especially homes, offices, businesses, shopping malls, factories and theaters worldwide.

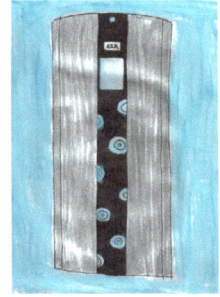

Willis Haviland Carrier, Father of Air Conditioner

HOW TO BECOME A MILLIONAIRE BY INVENTION?

100 Stories of Successful Inventions

Sunglasses

Sunglasses are a form of protective eyewear designed primarily to prevent bright sunlight and high-energy visible light from damaging or discomforting the eyes.

They can sometimes also function as a visual aid, as variously termed spectacles or glasses exist, featuring lenses that are colored, polarized or darkened. Sunglasses' usage is mandatory immediately after some surgical procedures, such as LASIK, and recommended for a certain time period in dusty areas, when leaving the house and in front of a TV screen or computer monitor after LASEK. Sunglasses have long been associated with celebrities and film actors primarily from a desire to mask their identity.

In prehistoric and historic time, Inuit peoples, indigenous peoples inhabiting the Arctic regions of Greenland, Canada and Alaska, wore flattened walrus ivory "glasses", looking through narrow slits to block harmful reflected rays of the sun. Also, it is said that the Roman emperor Nero liked to watch gladiator fights with emeralds. These, however, appear to have worked rather like mirrors. Sunglasses made from flat panes of smoky quartz, which offered no corrective powers but did protect the eyes from glare, were used in China in the 12th century or possibly earlier. Ancient documents describe the use of such crystal sunglasses by judges in ancient Chinese courts to conceal their facial expressions while questioning witnesses.

James Ayscough began experimenting with tinted lenses in spectacles around 1752. These were not "sunglasses" as that term is now used; Ayscough believed that blue- or green-tinted glass could correct for specific vision impairments. Protection from the Sun's rays was not a concern for him. Yellow/amber and brown-tinted spectacles was also a commonly prescribed item for people with syphilis in the 19th and early 20th centuries because sensitivity to light was one of the symptoms of the disease. Since the 1940s, sunglasses have been popular as a fashion accessory, especially on the beach.

LensCrafters, an American retailer of prescription eyewear and sunglasses, and the largest optical chain and the largest eyewear company in the U. S., developed "Transitions Vantage" lenses. Not only do they adapt to changing light conditions, they also polarize as they darken, reducing glare so the wearer can see life in crisp definition and vivid color outdoors.

Inuit snow goggles function by reducing exposure to sunlight, not by reducing its intensity.

HOW TO BECOME A MILLIONAIRE BY INVENTION?

100 Stories of Successful Inventions

Hula Hoop

In 1957, an Australian visiting California told Arthur K. "Spud" Melin and Richard Knerr Offhand that in his home country, children twirled bamboo hoops around their waists in gym class. Knerr and Melin saw how popular such a toy would be and soon they were winning rave reviews from school kids for the hollow plastic prototype they had created.

Arthur K. "Spud" Melin and Richard Knerr invented the modern Hula Hoop in 1958, but children and adults around the world have played with hoops, twirling, rolling and throwing them throughout history.

Hula Hoops for children generally measure approximately 71 centimeters (28 in) in diameter and for adults around 1.02 meters (40 in). Traditional materials for hoops include willow, rattan (a flexible and strong vine), grapevines and stiff grasses. Today, they are usually made of plastic tubing.

The Hula Hoop gained international popularity in the late 1950s, when California's Wham-O toy company successfully marketed a plastic version. In 1957, Richard Knerr and Arthur "Spud" Melin, started with the idea of Australian bamboo "exercise hoops", manufactured 1.06-metre (42 inch) hoops with Marlex plastic.

With giveaways and national marketing and retailing, a fad was started in July 1958, with twenty-five million plastic hoops sold in less than four months.

In two years, sales reached more than 100 million units.

Carlon Products Corporation was one of the first manufacturers of the Hula Hoop. During the 1950s, when the Hula Hoop craze swept the country, Carlon was producing more than 50,000 Hula Hoops per day. The hoop was inducted into the National Toy Hall of Fame at The Strong in Rochester, New York, in 1999.

The Hula Hoop craze swept the world, dying out again in the 1980s, but not in China and Russia, where traditional circuses and rhythmic gymnasts adopted hula hooping and hoop manipulation.

Arthur K. "Spud" Melin and Richard Knerr with Hula Hoops

HOW TO BECOME A MILLIONAIRE BY INVENTION?   79th Story / Total 100 Stories

100 Stories of Successful Inventions

JAVA

Nowadays, if you look at the classified ads for jobs, you can easily see the word "Java".

Java refers to the "Java" programming language associated with web pages. Even in our generation, our children have to know one or two programming languages before they want to get the job. What is "Java" language?

In the early '199s, Sun Company had a young, talented programmer named Patrick Norton.

"Is there any technology that enables communication between general household appliances or portable devices?" Under this motto, a programming language for creating applications for various devices and codenamed "Oak" is born. This language will later become "Java".

Today, it is often said that Java will play an important role in adding Internet-related functionality to cell phones and PDAs, as Java was the language that was created for Post PC appliances from the time of development.

Time Warner believed in Silicon Graphics' reputation and decided to work together. The need for a language that is strong on the network, platform independence, good stability, and security from viruses is more demanding than ever on the "Web". The Java development team has confirmed that all the requirements are met exactly enough to make sure that the web has emerged for Java, and completes the web browser based on Java.

Java technology can make the web an "interactive platform" and it is gaining global interest. It is true that Java has started with the overwhelming support of so many experts in the history of computers, but that's because the performance of Java and the growth potential of Java are enough to meet that expectation.

Just as the web browser has become an international necessity, Java, the language of the web browser, has also spread to the whole world.

With 'Java" web language, we can communicate with each other through the Internet. Also, general household appliances or portable devices can communicate each other.

HOW TO BECOME A MILLIONAIRE BY INVENTION?   80th Story / Total 100 Stories

100 Stories of Successful Inventions

Cooking Help App

It is said that man should listen to the words of three women for his lifetime: mother, wife, and car navigator assistant woman ... Now it seems that man should listen to the chef's assistant lady, too.

This is because Pantelligent, developed by MIT engineer Humberto Evans, is gaining popularity.

With the app dedicated to smart phone apps, you choose whichever recipe you want to create. As well as showing how to make and materials, you can manage the temperature and time during cooking. If you are staring at a smart phone screen rather than food, you do not have to think about when you need to put in any other ingredients or add another ingredient. In the first few minutes of cooking, you know what stuff you need to put in the bluetooth and tell it on your smart phone.

It is a smart frying pan that helps cooking carefully by displaying temperature and cooking process on smart phone or tablet. It has a built-in Bluetooth LE wireless communication device on the handle and displays information wirelessly transmitted to the smart phone using a frying pan thermometer sensor.

The voice guidance tells you when you need to cook and when you need to mix or turn over, so that you can complete a delicious dish. Frying pans and smart phones can make the best dishes as they say and do not have to worry about picking up food anymore to control the temperature of the food.

Precise temperature control is the core technology of this Pantelligent. In order to follow the cooking method of the best hotel chef, every one of the cooking methods is informed, and the chef of the smart phone follows the instructions of the chef App, and a fantastic dish is made. When the dish is finished, the alarm sounds, and good recipes can be shared among friends in the dedicated apps SNS community.

This frying pan makes you a great chef when you listen to the chef lady of smart phone app.

HOW TO BECOME A MILLIONAIRE BY INVENTION? 81st Story / Total 100 Stories

100 Stories of Successful Inventions

Ringer's Solution

When a child is born, he sucks a bottle of milk. When he is a child, he drinks cola with pizza. When he becomes an adult, he drinks beer. When he becomes old and gets sick, he might need Ringer's solution in his bed.

What is Ringer's Solution? Ringer's solution is a solution of several salts dissolved in water for the purpose of creating an isotonic solution relative to the body fluids of an animal. Ringer's solution typically contains sodium chloride, potassium chloride, calcium chloride and sodium bicarbonate, with the last used to balance the pH. Other additions can include chemical fuel sources for cells, including ATP and dextrose, as well as antibiotics and antifungals.

Ringer's solution is frequently used in in-vitro experiments on organs or tissues, such as in vitro muscle testing. The precise mix of ions can vary depending upon the taxon, with different recipes for birds, mammals, freshwater fish, marine fish, etc. It may also be used for therapeutic purposes, such as arthroscopic lavage in the case of septic arthritis.

Ringer's solution is named after Sydney Ringer, who in 1882–1885 determined that a solution perfusing a frog's heart must contain sodium, potassium and calcium salts in a definite proportion if the heart is to be kept beating for long.

This solution was adjusted further in the 1930s with the addition of sodium lactate forming Ringer's lactate solution.

Ringer's lactate solution (RL), also known as sodium lactate solution and Hartmann's solution, is a mixture of sodium chloride, sodium lactate, potassium chloride, and calcium chloride in water. It is used for replacing fluids and electrolytes in those who have low blood volume or low blood pressure. It may also be used to treat metabolic acidosisin cases other than those caused by lactic acidosis and to wash the eye following a chemical burn. It is given by injection into a vein or applied to the affected area.

It is on the World Health Organization's List of Essential Medicines, the most effective and safe medicines needed in a health system. It is used for replacing fluids and electrolytes in those who have low blood volume or low blood pressure.

You might need Ringer's Solution if you have low blood volume or pressure.

HOW TO BECOME A MILLIONAIRE BY INVENTION?

100 Stories of Successful Inventions

Sovaldi

Hepatitis C is a disease that affects the liver. It is caused by infection with the Hepatitis C virus, which is spread through contact with the blood of an infected person. For most people, Hepatitis C will become a chronic infection, which means that the virus stays in the body for many years. Chronic Hepatitis C can eventually lead to serious liver damage, liver failure, or liver cancer.

Hepatitis C cannot be prevented with a vaccine, and once it is contracted, it is more than 70% likely to become chronic and develop into cirrhosis or liver cancer. Although it is possible to manage with medication, the cost of treatment is considerable and drug-related side effects should also be considered. There is a medicine called Sovaldi made by Gilead. It is more effective than the existing hepatitis injections, has fewer side effects, and is a Hepatitis C drug that can be taken orally instead of via injection and has been attracting attention because it is easy to take. But the price of the medicine is the beyond the reach of ordinary people. The drug is sold in the United States for $ 1000 a month and $ 12,000 for $ 84,000. However, it has a cure rate of over 90% for 12 weeks of treatment. Sovaldi is a 'blockbuster' drug that has been sold more than billions of dollars annually since its launch in 2013.

The success of the Sovaldi is based on a change in the paradigm of the drug. Meanwhile, chronic hepatitis was treated based on peg interferon.

In particular, combination therapy with ribavirin improved the outcome of chronic hepatitis but still failed in 30-50% hepatitis patients and was limited in patients with advanced liver disease due to toxicity. Recently, Gilead's company has developed a direct-acting antiviral agent (DAA) that directly acts on the virus target, thus improving the success rate of chronic HCV treatment and anticipating it to replace antiviral therapy based on peg interferon. The Approximate 150 million people with chronic Hepatitis C are estimated to be around the world, and are mostly likely from low or middle-income countries. Therefore, the process of gaining access to this medicine is receiving worldwide attention.

Fortunately, Gilead and several other pharmaceutical companies are now preparing generic drugs for Sovaldi.

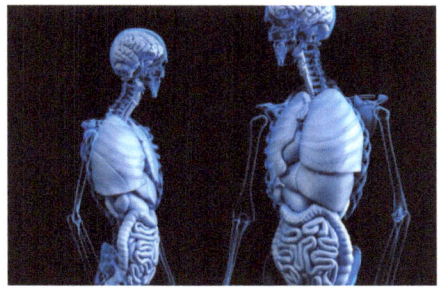

Some diseases like Hepatitis C cannot be prevented by vaccines, thus the invention of medicines like Sovaldi is very important.

HOW TO BECOME A MILLIONAIRE BY INVENTION?

100 Stories of Successful Inventions

Panera Bread

My favorite place for breakfast is Panera: fresh bagels, tasty sandwiches, hot coffee with free refills, delicious soup, cozy atmosphere, etc.

Panera Bread that operates more than 2,000 stores across the United States and Canada. Even when real estate plunged in 2008, Panera increased the number of stores. Panera was founded by Ronald Shaich and others more than 30 years ago in 1987, and the founder and current chairman of Ronald Shaich's philosophy is that "if there is a viable concept, it is the best time for business expansion and growth even during the worst recession. As we expand our business, we are always developing new menus such as salad, macaroni, soup, and improving quality, and there is also a royalty program for enthusiastic customers."

Panera products are known to use whole grain bread, chicken treated without antibiotics, eco-friendly materials, and no trans fat as well as having low salt and low-calorie options such as its chicken soup.

Also, if a new menu item is developed, customers will be offered free tasting opportunity to develop active marketing. Not only has it actively secured breakfast, lunch, and dinner guests, but also provides catering services to business customers such as companies.

Catering services have grown by more than 20% each year since 2010.

Other restaurants in the economic depression period were trying to keep costs and minimize employees, but Panera maintains the quality of food and free refill coffee and beverages, and continued to improve the menu and customer service while expanding the business.

Panera Bread provides breakfast, lunch and dinner in cozy atmosphere and also offers convenient catering services.

HOW TO BECOME A MILLIONAIRE BY INVENTION?

100 Stories of Successful Inventions

Sulwhasoo Cosmetics

Korean Red Ginseng is famous for its effectiveness.

Sulwhasoo, being made by Korean Cosmetics Amorepacific Company, uses Korean Red Ginseng as one ingredient. At the heart of Sulwhasoo is Korean herbal medicine, steeped in a long tradition of Eastern holistic philosophy that searches for deeper answers, with a desire to find harmony and balance, and treatments that view the body as a whole.

Sulwhasoo originator Suh Sung-Whan spent his childhood in Kaesong, a city renowned for the meticulousness of its merchants and the exceptional quality of its ginseng. His mother produced camellia oil there and became his lifelong mentor. Under the watchful eye of his mother and her uncompromising devotion to product quality, Suh learned the value of nature and the importance of high quality ingredients. Determined to make cosmetics using his childhood learning and the efficacies of ginseng, Korea's signature plant, Suh introduced Ginseng Cream in 1966.

In 1972, after nine years of ginseng research, Suh succeeded in extracting ginseng's most precious saponin from the plant's leaves and flower. In the following year, Ginseng SAMMI with ginseng saponin as the key ingredient was introduced. Ginseng SAMMI represented Suh's dream to introduce a cosmetic embodying Korea's culture and ingredients to the world, complete with a specialized container inspired by the exquisitely crafted Goryeo celadon.

Sulwha, a gentle cosmetic product that used natural Korean herbal ingredients to help protect the skin and give it strength, was created in 1987, referring to a beautiful snow flower that blooms to herald the coming of spring, just as beauty blooms from a woman's skin. The brand evolved into the Sulwhasoo skincare brand in 1997 with Suh's deep conviction in the power of ginseng and his passion for Korean herbs.

Today, it is beloved by customers worldwide as the top beauty product brand due to natural ingredients like Korean ginseng and unique marketing.

Sulwhasoo Cosmetics has become popular using Korean Ginseng as one ingredient.

HOW TO BECOME A MILLIONAIRE BY INVENTION?   85th Story / Total 100 Stories

100 Stories of Successful Inventions

Artificial Heart

Worldwide, more than 2 million people suffer from heart disease, and in the United States alone, there are about 5 million people with heart disease. But every year there are approximately 2,500 cardiac donors.

In 1969, Dr. Denton Cooley of the Texas Heart Institute in the United States established the first artificial heart in the world as a connection to heart transplantation for a desperate heart patient.

On December 1, 1982, the researchers at the University of Utah began the artificial heart permanent implantation procedure for the human body. The "Zabic 7" type artificial heart used in the surgery was developed by Dr. Zabik and consists of two artificial ventricles corresponding to the left and right ventricles of the heart. In the real heart, only the left and right ventricles are removed, and instead the artificial ventricles are connected to the atria and the aorta and the pulmonary artery. Unfortunately, he died 112 days after surgery, due to blood clots and bacterial infections. This device also has a large external power source that is inconvenient to move.

In 2001, it was the first time in the United States that artificial heart transplantation was able to completely replace human sick heart function. The Louisville University Medical Center in Kentucky succeeded in transplanting an artificial heart at the end of the seven-hour operation at the Juice City Hospital.

The artificial heart, named "AbioCor", weighs 900 grams, made by the American company Abiomed and it fits inside the human body. "This device is comfortable enough to live with and forget about after implantation," said Dr. Metz Oz, director of the Cardiovascular Research Institute at Columbia University. Abioco Heart is designed to stimulate the heart rhythm as blood flows through the lungs to other parts of the body. This device with two chambers has a motorized hydraulic pump system. AvioCor Heart is operated by an internal battery that is constantly recharged by the energy of the external battery. Patients wear an external battery at the waist, and the energy conversion device attached to the skin transfers the energy of the external battery to the internal battery. AvioCor artificial heart has changed from the existing 'air driven' to 'electric driven', which allows the patients to move out of the room, giving heart patients a lot of hope.

Early Model of Artificial Heart Displayed in London Museum

HOW TO BECOME A MILLIONAIRE BY INVENTION?

100 Stories of Successful Inventions

86th Story / Total 100 Stories

Self-Driving Car

When I drive a car during rush-hour traffic, I hope my car does driving for me. My wish might come true soon. Tesla, Google, Volvo, Toyota, Hyundai, Kia, and other companies are working to develop autonomous vehicles that are already running without a driver.

Recently, Audi developed the RS7 autonomous drive car in 17 touring car masters race. It is said that it completed 4574km at the maximum of 240km / h in 2 minutes and 10 seconds. It is faster than a few racers. The RS7 intuitively ran along the correct racing line without delay, grabbed the turn-in point, and read the track in cm increments. Like a seasoned racer, the car passed the Apex and delayed braking when the car entered the corner. The steering control algorithm was a bit smoother and the car decided to accelerate when the car got out of the cornering section, but it was amazingly skillful overall. According to test records measured by Audi engineers, the gravitational acceleration of the RS7 unmanned autonomous drive system at braking was 1.3 g and 1.1 g at cornering. The RS7 also incorporates the latest GPS technology to keep track on the track. Using this technology, data calculated to the nearest centimeter is transmitted to the car through the wireless communication network, and the 3D image collected from the camera installed on the windshield is compared with the graphic image stored in the system. The system analyzes countless amounts of images to identify hundreds of distinctive features and then uses them as additional location information.

It is said that it might take more than 10 years to mass-produce self-driving cars because the legal requirements of safety requirements. It looks like law and society are more roadblocks for self-driving car than current technology.

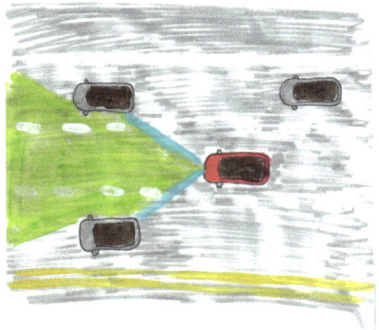

Carmakers foresee a future of self-driving cars truly autonomous vehicles guided by high definition maps, GPS Navigator and an array of cameras, sensors and scanners.

HOW TO BECOME A MILLIONAIRE BY INVENTION?   87th Story / Total 100 Stories

100 Stories of Successful Inventions   Watson (Artificial Intelligence) by IBM

Google surprised the world by introducing "AlphaGo" Artificial Intelligence computer program, which became the champion in Go board game in 2015.

IBM's Artificial Intelligence project is called Watson. Watson is a question answering computer system capable of answering questions posed in natural language, developed in IBM's DeepQA project by a research team led by principal investigator David Ferrucci. Watson was named after IBM's first CEO, industrialist Thomas J. Watson.

The computer system was specifically developed to answer questions on the quiz show 'Jeopardy' and in 2011, the Watson computer system competed on 'Jeopardy' against former winners, Brad Rutter and Ken Jennings and Watson won the first place prize of $1 million.

Watson had access to 200 million pages of structured and unstructured content consuming four terabytes of disk storage [1]including the full text of Wikipedia, but was not connected to the Internet during the game. For each clue, Watson's three most probable responses were displayed on the television screen. Watson consistently outperformed its human opponents on the game's signaling device, but had trouble in a few categories, notably those having short clues containing only a few words.

In 2013, IBM announced that Watson software system's first commercial application would be for utilization management decisions in lung cancer treatment at Memorial Sloan Kettering Cancer Center in New York City, in conjunction with health insurance company WellPoint.

IBM Watson's former business chief, Manoj Saxena, says that 90% of nurses in the field who use Watson now follow its guidance.

Artificial Intelligence is getting better every day and AI would be more important in the future.

HOW TO BECOME A MILLIONAIRE BY INVENTION?

100 Stories of Successful Inventions

Yamaha Clavinova

There are many parents who sign up their children for piano and violin lessons. Learning how to play the piano or violin instrument from a young age has many advantages, including improvement in speaking and reading, as well the ability to improve concentration and reduce the problematic behavior. In addition, life tends to become richer with knowledge of music, but there are too many instruments and too little time for a child.

The Clavinova is a long-running line of premium digital pianos created by the Yamaha Corporation.

They are similar in styling to an acoustic piano, but with many features common to various keyboards such as the ability to save and load songs, the availability of different voices, and, in more recent models, the ability to be connected to a computer via USB or wireless network for music production or interactive piano lesson programs.

The built-in synthesizer produces the sound. Early Clavinova models used FM synthesis, but later models use sample-based synthesis to produce the sound. Information comes in a MIDI or similar format either directly from the piano keyboard or from a stored source (from within the piano or via a computer or external sequencer).

The synthesizer can imitate a large array of acoustic instruments, electronic instruments and other sound effects.

Recent models of CVP Clavinova have hundreds of such voices. These usually include many types of pianos and organs, string, percussion, brass and woodwind instruments, as well as modern and vintage synthesizer sounds, sampled effects etc.

The more recent CVP models also feature many accompaniment styles, ranging from traditional dance and classical orchestration, through to more modern club, pop, rock, big band and jazz styles.

Yamaha Clavinova has many sound effect options: piano and organ, string, percussion, brass and woodwind instruments.

HOW TO BECOME A MILLIONAIRE BY INVENTION?

100 Stories of Successful Inventions

Apple Pencil

Nowadays, with the popularity of smart devices and computers with keyboards and mouse devices, we are using less pencils, ballpoint pens, and brushes. Furthermore, tablets with touchscreens have made pencils, ballpoint pens and brushes obsolete.

Apple recently developed the Apple pencil, a touch pen to replace pencils and drawing pens. Whether you sketch in the park, paint a watercolor portrait, draft a blueprint, or just have an apple pencil, you have all the tools you need. There is only one tool in your hand, but there is no limit to what you can do. Apple pencils distinguish themselves from other creative devices because of their fast response speed. There is little delay between the moment you start drawing and the moment you draw on the screen. When IPad Pro detects an Apple pencil, the subsystem scans the Apple pencil's signal at a rate of 240 times per second. It is twice the amount of point data collected from the finger. With the data collected and the software designed by Apple, it takes only a few milliseconds for the images to be drawn into the display to be displayed on the display. In a sleek, small Apple pencil enclosure, complex high-precision pressure sensors are built in to measure a wide range of pressures. This carefully placed sensor accurately detects how much force is applied to the tip of the Apple pencil. As more pressure is applied, a thicker line is drawn, and a lightly pressed line is drawn like a hair. Through this, the way to express creativity becomes infinite. Two tilt sensors built into the Apple pencil tip accurately measure the direction and angle of the hand. When a person writes or draws a picture just like usual,

Multi-Touch display detects the relative position of each sensor. Similar to the charcoal pencil or a normal pencil, the Apple pencil can be tilted to add a shading effect. The IPad Pro has a pampletion feature and Apple pencil can be used even with a hand resting on the screen, so people can continue drawing without worrying about arm position.

The lightning and charge cable is designed to be slightly longer, allowing charging even when the IPad Pro is in the silicone case. When fully charged, people can spend 12 hours on graffiti, sketching, annotating, editing, and artists have become the age of making artworks with canvas, brushes, and paint, not paint.

Apple Pencil allows for far finer control of hand drawn notes, figures, etc.

HOW TO BECOME A MILLIONAIRE BY INVENTION? 90th Story / Total 100 Stories

100 Stories of Successful Inventions

Samsung Galaxy Gear VR

In a science fiction film called "Total Recall," there are scenes in which people travel through virtual reality in the future. In the movie, people do not book a flight ticket or hotel, but they book the virtual trip. It is popular because people lie still in the room, and the virtual trip offers cheap, more realistic, all-round experience. Workers living day by day can travel virtually anywhere to places as far as Mars.

Recently, Samsung released a new model of gear VR linked with the Galaxy Note. With the new gear VR, people can use more than 300 virtual reality applications such as 360-degree videos, games, education, and more than 100 apps such as games and videos recommended by the Oculus Store. With Gear VR, people are able to experience virtual reality as if they are present at incredible games, world famous natural landscapes, historical places, and music performances, roller coaster rides, virtual experiences, and so on. With a wide viewing angle of 96°, it improves immersive and realistic senses. It senses head movement in real time with precise head tracking, so you can experience a 360° panoramic view that changes according to the movement of your eyes in any direction. Gear VR is also compatible with various Samsung Galaxy Smart phones. Obama, former President of the United States, used the new Virtual Reality (VR) headset in the office of the White House and looked up at the ceiling.

The VR worn by President Obama is 'Gear VR' jointly developed by Samsung Electronics and Facebook subsidiary Oculus.

When the White House released the picture, Mark Zuckerberg, the CEO of Facebook, wrote on his Facebook page, "Let's have a gravity table tennis next time!" meaning that he wanted to experience real table tennis in the air while watching VR video. With all of this technological and virtual reality advancement, it seems like The 'Matrix' movie, which makes people feel the virtual reality as real reality by controlling the five senses, might come true in someday. The future with advanced virtual reality technology may become very interesting.

President Obama is experiencing Samsung "Gear VR."

HOW TO BECOME A MILLIONAIRE BY INVENTION? 91st Story / Total 100 Stories

100 Stories of Successful Inventions

Fuji Cosmetics

Once upon a time, before the cell phone camera became popular, before the digital camera came out, people used cameras that required inserting rolls of film. At that time, there were two powerful companies that provided this film: Kodak and Fujifilm. At first, people were only able to take 24 pictures, but later people could take more pictures up to 36.

Nowadays, using cell phone cameras, we can take thousands of pictures or video. The Kodak company, which monopolized film and cameras, disappeared from history as it went bankrupt in 2012, but what about Kodak's competitor, Fujifilm?

Surprisingly, Fujifilm, which was a film company, is now a cosmetics company. It also handles medical device products and health foods.

What caused these amazing changes in Fujifilm? Fuji succeeded in developing a cosmetic product called Astalift by applying collagen control technology obtained from developing film. As a result, Fuji succeeded in making up 41% of sales of cosmetics and health care products. An important turning point is the difference in how to cope with the rapid decline in film-related sales, which began in 2000. Fujifilm has restructured its related businesses as film-related businesses have shrunk, rather than solely reducing R & D manpower. R & D research strategy to recycle existing technology to other fields is key, as constant self-innovation efforts are important. IT companies have invested heavily in R & D investment. With the strong slogan of the second start-up, it has begun to develop new business technology while applying existing base technology for a new product. It is time for other companies to invest in R & D development, and to anticipate future demand like Fujifilm, which has survived by adapting to the continuously changing industry.

Fujifilm, a film company, with collagen technology, has developed Astalift cosmetics.

HOW TO BECOME A MILLIONAIRE BY INVENTION?

100 Stories of Successful Inventions

92nd Story / Total 100 Stories

Microsoft Office

Bill Gates once said, "I choose a lazy person to do a hard job. Because a lazy person will find an easy way to do it." Bill Gates, who has always been among the world's 10 richest people, founded Microsoft in 1975.

To make life easier, people invented a typewriter in 1874. A typewriter is a mechanical or electromechanical machine for writing characters by means of keys that strike a ribbon to transmit ink or carbon impressions onto paper.

However, to make life easier and easier, people invented word processor, which is an electronic device or computer software application that performs the task of composing, editing, formatting, and printing of documents.

The Microsoft Word Processor program is a word processing program and it is being used by over 90% of the world. Also, Microsoft OS program like Windows is running in most new PCs shipped. Microsoft Office makes modern office works easy by providing word processor, spreadsheet, database, presentation, email, etc.

Thus, Office software is mandatory for companies and it would not be easy to bypass Microsoft products like death and tax.

Recently, Microsoft released Office 365, a software purchasing policy that allows people to use it for one year or one month. Microsoft Office 365 includes Office Professional Plus for productivity and collaboration.

You can use Exchange Online, SharePoint online, and Lync Online to make payments on a monthly basis. With Office 365, people can work anywhere using PCs, smart phones, and tablets. It also provides cloud-based backup management tools that are easy to use.

It is estimated that 1.2 billion people are using Microsoft Office products.

HOW TO BECOME A MILLIONAIRE BY INVENTION?

100 Stories of Successful Inventions

PROBIOTICS

When I was young, eating yogurt was enough to maintain the stomach health. These days, many people are taking Probiotics to maintain the stomach health.

What are Probiotics and are they really good for your stomach health?

Probiotics are defined as live microorganisms that are believed to provide health benefits when consumed. The term probiotic is currently used to name ingested microorganisms associated with benefits for humans and animals. The term came into more common use after 1980.

The introduction of the concept (but not the term) is generally attributed to Nobel laureate Élie Metchnikoff, who postulated that yogurt-consuming Bulgarian peasants lived longer lives because of this custom. He said in 1907, "the dependence of the intestinal microbes on the food makes it possible to adopt measures to modify the flora in our bodies and to replace the harmful microbes by useful microbes".

A significant expansion of the potential market for Probiotics has led to higher requirements for scientific substantiation of putative benefits conferred by the microorganisms.

Although there are numerous claimed benefits of using commercial Probiotics, such as reducing gastrointestinal discomfort, improving immune health, relieving constipation, or avoiding the common cold, such claims are not backed by scientific evidence and are prevented as deceptive advertisements in the United States by the Federal Trade Commission.

While eating yogurt was more fun, it might be cost effective to take Probiotics capsule with billions of good bacteria in it.

Probiotics with billions of good bacteria

HOW TO BECOME A MILLIONAIRE BY INVENTION?

100 Stories of Successful Inventions

94th Story / Total 100 Stories

Fidget Toys

When I was a student, I used to spin a ballpoint pen or a pencil maybe because of too much stress for school exams.

Students who are suffering from continuing stress of examinations and adults who are constantly stressed in the workplace and in business life need some kind of relaxing toys.

The toy 'Fidget Cube', a toy aimed at modern people with an uneasy psychology, hit the mark. Kick-starter, the world's largest crowd funding site, has exceeded its goal of $ 15,000, exceeding $ 4.58 million. Since starting funding, the number of investors has risen to 112,200, the second highest in Kick-starter history. The 'desk toy', which means toys to play on the desk, is a simple function weighting just 40g with a small size of 3.3cm. Each side of the cube has a different function button that allows you to repeat simple actions such as pressing, turning or rubbing your fingers. It is aimed at the habit of people who do not touch the ring on their fingers or tickle the ballpoint pen, help relieve stress, relax and increase concentration. It is analyzed that it receives the enthusiastic support of workers all over the world by eliminating the desire to be irritated or bored. Also, Fidget Spinner, a toy aimed at modern people, are popular, too and Fidget Spinners are doing nothing but spinning, but it might relieve stress, relax and increase concentration for stressful adults and students.

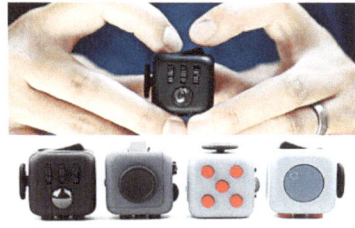

Fidget Toys for Stressed People

HOW TO BECOME A MILLIONAIRE BY INVENTION?

100 Stories of Successful Inventions

Canned Food

In the history of mankind, people seem to have always waged war. So, in order to win the battle against the enemy, a number of weapons and other novel inventions were made. But another invention made for war is one that we see in our homes every day: canned food.

Nicolas Appert invented canned products in 1804. The way he devised was a simple way of putting food in a wide glass bottle, immersing it in boiling hot water, and sealing it tightly with a cork when the contents were hot. Nicholas Appert's motive was to recruit the French Emperor Napoleon I, who had already conquered Europe in 1795, to find a way to completely save the food supplied to the army in order to conquer Russia in the future. Inspired by this, Nicholas Appert developed his own past experience of managing a food factory, a wine factory, and putting food into a glass bottle, sealing it and storing it. He announced this in 1809 and was awarded 12,000 francs by the French government.

Then, in 1810, Peter Durand of England devised a method of using tin containers made of metal, instead of glass bottles. He called it Tin Canister.

Since the first can invention by Appert and the metal can invention by Durand, the can manufacturing technology has greatly advanced in the United States. The manufacturing of canned tin pipes was gradually mechanized and the technology was changed.

During the Second World War, about two-thirds of the US and other allied food was supplied as canned food. Although canned food was invented as a war material to be used for war, canned food is also used as a relief and humanitarian material for people suffering from food shortages such as poverty, earthquake and flood.

Nicolas Appert used glass bottles to store food for the army during war.

HOW TO BECOME A MILLIONAIRE BY INVENTION?

100 Stories of Successful Inventions

Rain-X

In Greek mythology, Icarus flew into the sky with wings made of feathers, imitating a bird flying freely in the sky. But Icarus rose so high that he paid for his mistakes, as the bee wax that was used to stick the feathers melted and fell into the sea and Icarus fell down to the earth. Though the mythical 'Wings of Icarus' ended in tragedy, people have imitated nature for a long time. It is not only biological reactions of animals, plants, or microorganisms, but also observation and imitation of various natural phenomena to create many inventions and to develop science and technology.

On the surface of the lotus leaf, there are fine protrusions, so that even when it rains, water does not penetrate. Instead, water droplets are formed and flow down along the surface. At this time, the dust on the surface of the leaf rolls together. The self-cleaning function, which keeps itself clean even if it is not washed like a lotus leaf, is called "lotus leaf effect". In the meantime, many scientists have been studying to make materials that have lotus leaf effect. Some inventions with lotus leaf effect have already been used in real life. An example would be a lotus leaf that flows down the surface without spilling the juice or ketchup. However, in the meantime, there was no cloth to show the effect of the lotus leaf in hot water.

Not long ago, the University of Minnesota and the Hong Kong Science and Technology Institute collaborated in inventing a new lotus leaf texture that does not get touched by hot coffee or milk. There is still a problem to be solved, such as the color and texture of the fabric, but it must be a valuable invention that makes life convenient.

On the other hand, there is a product called Rain-X, which is an automotive water-repellent glass coating agent mimicking this lotus leaf effect. Normally, when you put on a glass window through a paper towel or sunlight, you can see miraculous images of raindrops rolling down when it rains. If you observe plants, insects and animals in nature as well as daily life with the eye of curiosity, you can get the inspiration and wisdom of invention that can be a great hit product in the market.

The "lotus leaf effect" inspired many inventions. One of them is Rain-X.

HOW TO BECOME A MILLIONAIRE BY INVENTION?

97th Story / Total 100 Stories

100 Stories of Successful Inventions

3D Printer for Food

Not only in the United States but also in other developed countries, the proportion of people who live alone has increased, exceeding 27 percent this year. Although right now the number of these households remains at 25 percent, it will be over 34 percent in approximately 20 years.

At that time, one in three family would be one-person family. The most serious of the problems caused by the increase of one person in culture is the collapse of food culture. Preparing food for the whole family would be uncommon, and the dining culture that we eat together happily disappears. For those who live alone, meals are often synonymous with fast food like hamburgers and pizza and home-heated prepared meals.

Nowadays 3D printing is also entering the cooking market. A 3D printer, Foodini, which makes food from computers and food databases, has already been released. By using 3D printing, you can produce healthy foods that can be eaten directly and can be eaten right away, so technology can produce very healthy and fresh food. 3D printing food can be said to be sustainable because it shortens the production chain of food production, reduces transportation costs, and requires less land for cultivation and production. Also, it would be good news for people who are suffering from weight loss or diseases that require dietary control.

3D printing food accurately calculates the calories and also learns each individual's needs and tastes, the food is made from the data that is available. It would not be long before I could eat a 3D printer's food every day, even if I was not a professional chef. 3D printing will provide a form that was previously impossible and proven tasty and delicious menu. It is made using a capsule of ingredients that prints a layer of a specific food type, and the printing device can be connected to the Internet to constantly update a variety of new but verified recipes or materials. In addition to food, all the furniture and tools used for meals will be made to be 3D-printed. Using online data, the robot will analyze the atmosphere, nutrition and taste of the participant's individual, and offer a menu corresponding to it, which will communicate with people and serve the food made with 3D printing. You can eat delicious, healthy and nutritious food by 3D printing in future.

Soon, 3D Printing will print food for you and your family.

HOW TO BECOME A MILLIONAIRE BY INVENTION?   98th Story / Total 100 Stories

100 Stories of Successful Inventions

Perfume

Perfume is a fragrance that increases people's mood and gives people a particularly good feeling. From when did perfume appear in human history?

The use of perfume by humans in their lives dates back to 4,500 years from now (as shown in the picture).

In the 9th century the Arab chemist Al-Kindi wrote the Book of the Chemistry of Perfume and Distillations, which contained more than a hundred recipes for fragrant oils, salves, aromatic waters, and substitutes or imitations of costly drugs. The book also described 107 methods and recipes for perfume-making and perfume-making equipment, such as the alembic for distilling chemicals.

The Persian chemist Ibn Sina introduced the process of extracting oils from flowers by means of distillation, the procedure most commonly used today. He first experimented with the rose. Until his discovery, liquid perfumes consisted of mixtures of oil and crushed herbs or petals, which made a strong blend. Rose water was more delicate, and immediately became popular.

Also, the art of perfumery was known in Western Europe from 1221, taking into account the monks' recipes of Santa Maria Novella of Florence, Italy.

In the east, the Hungarians produced in 1370 a perfume made of scented oils blended in an alcohol solution – known as Hungary Water – at the behest of Queen Elizabeth.

The art of perfumery prospered in Renaissance Italy, and in the 16th century Rene the Florentine took Italian refinements to France. Thanks to Rene, France quickly became one of the European centers of perfume and cosmetics manufacture.

Between the 16th and 17th centuries, perfumes were used primarily by the wealthy to mask body odors resulting from infrequent bathing. Partly due to this patronage, the perfume industry developed. In 1693, Italian barber Giovanni Paolo Feminis created perfume water called Aqua Admirabilis, today best known as eau de cologne. By the 18th century and even today, Italy and France remain the center of European perfume design and trade.

Egyptian scene depicting the preparation of lily perfume, 4th century BC

Perfumes, used in religious ceremonies since early civilization, can be found all around the world.

HOW TO BECOME A MILLIONAIRE BY INVENTION? 99th Story / Total 100 Stories

100 Stories of Successful Inventions

Hangul (or Korean Alphabets)

Do you know that most people can learn Korean Alphabets (24 alphabets with 14 consonants and 10 vowels only) and read Korean in a couple of days?

Korean Alphabets or Hangul were designed so that even a commoner could learn to read and write; the alphabet book says, "A wise man can acquaint himself with them before the morning is over; a stupid man can learn them in the space of ten days."

In 1446, Sejong the Great, the fourth king of the Joseon Dynasty, promulgated Hangul Alphabet. HunMinJeongEum HaeRye ("Korean Alphabet Explanation and Examples") explains the design of the consonant letters according to articulatory phonetics and the vowel letters according to the principles of vowel harmony. Before the creation of Hangul, people in Korea (or Joseon Dynasty) primarily wrote using Classical Chinese characters. However, due to the fundamental differences between the Korean and Chinese languages with up to 50,000 characters, and the large number of characters needed to be learned, there was much difficulty in learning how to write using Chinese characters for the lower classes, who often didn't have the privilege of education. To assuage this problem, King Sejong created the unique alphabet known as Hangul to promote literacy among the common people.

In its original forms, the alphabet has 19 consonant and 21 vowel letters. However, instead of being written sequentially like the letters of the Latin script, Hangul letters are grouped into blocks and each of which transcribes a syllable. That is, although the syllable may look like a single character, it might be actually composed of two or more letters including at least one consonant and one vowel. These blocks are then arranged horizontally from left to right or vertically from top to bottom. Each Korean word consists of one or more syllables, hence one or more blocks. The number of mathematically possible distinct blocks is 11,172.

Language makes socially weak people read, think, speak, and write to express their thoughts and make them strong socially. Hangul is such a great character and invention. By allowing people to learn easily, to express their thoughts freely, and to be adaptable in computer keyboard (and computer age) with just 24 alphabets, Hangul is one of the greatest inventions.

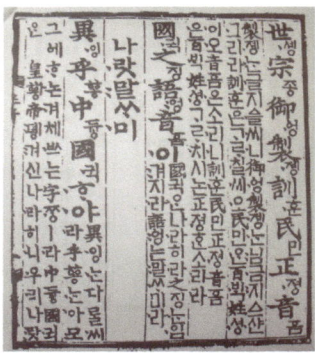

Korean Alphabet, invented by King Sejong 500 Years Ago.

HOW TO BECOME A MILLIONAIRE BY INVENTION? 100th Story / Total 100 Stories

100 Stories of Successful Inventions

Robot Vacuum and Robot Dog Walker

In one Korean drama, there was a scene in which a mermaid lady was afraid of an automatic vacuum cleaner. It was funny, but this automatic vacuum cleaner did not exist even just a decade ago and I would have been afraid to see the robot vacuum, too.

However, in many households, vacuum cleaners have become an essential necessity. The reason people use automatic vacuum cleaners is because they are quiet, clean and comfortable. Once you put it on the charger, the robot vacuum sweeps the carpets and floors once or twice a day, cleans the floor with no noise. There is no inconvenience to use it because it is silent while you are watching favorite TV. So there are a lot of houses that have one or two robot vacuums in the family. An automatic vacuum cleaner is an example of a stunning invention using the psychology of people who are annoyed by house cleaning or vacuuming. On the other hand, those who raise puppies are supposed to walk them every day. I would not mind because of my health, but I would feel annoyed if I have to do that every day.

Now. RoboDynamics Company in Santa Monica, USA, introduced a robot, which can walk the dog. The name of the robot comes from the goddess Luna of the Moon. Now, Android robots, which are similar to people, are expected to come closer to the public with low prices and various functions. Luna can exercise pets, bring drinks to the owner, and help the doctors and nurses if they are used in the hospital. RoboDynamics is confident that Luna will revolutionize the robotics sector, just as PCs have revolutionized home computing and smart phones have revolutionized mobile devices. RoboDynamics is currently raising funds at Kickstarter, an Internet social funding site, to fund facility funding to build Luna's mass production system. Early adopters who are aware of the true value of Luna and are making early purchases, you can make an order reservation, too.

Luna is five feet tall and has arms long, HD camera, LCD, microphone and so on. Rolling through the wheel. The nature of ordinary people pursues ease and comfort, and inventions that make use of this nature are a big hit. The standing man wants to sit, and the sitting man wants to lie down. It seems that our future would be increasingly dominated by the world, such as cleaning robots, washing robots, robots that let dogs walk, and robots doing annoying things for those who are looking for this easy life. It would be necessary to use industrial robots to replace people's hard work at factories. However, warfare robots in the movie 'Terminator' and 'Star Wars' gives people anxiety about the future because these war robots can threaten people's lives and industrial robots might take away the jobs of people who have to make money for a living.

Automatic Vacuum, a necessary item in the modern home

Personal Robot by RoboDynamics, Luna, with the capability of walking a dog

HOW TO BECOME A MILLIONAIRE BY INVENTION?

100 Stories of Successful Inventions

CONCLUSION

And now, the 100 Stories of Successful Inventions has come to a close, but their success Stories and the secrets behind them continue to be big hits.

"Those who do not learn history are doomed to repeat it."

I would like to think that if we learn the secrets of success, we could repeat the success and we could invent an important invention, which would make people's life easier.

The Greek god of opportunity, Caerus, is said to have thick hair on the front of his head, symbolizing the option to be able to 'grasp' him as he approaches, yet the back of his head is bald, as he is unable to be grasped once he leaves.

Believe in yourself and you would have many great new or improvement ideas in your life.

If so, do not forget to write down the ideas in the next three pages!

You could have new ideas with flash of genius, but most likely you could try to modify the current invention, to improve the current invention by adding a new function, to combine two inventions into one or to use the current invention for another purpose.

Do not let that opportunity or your great ideas pass you by.

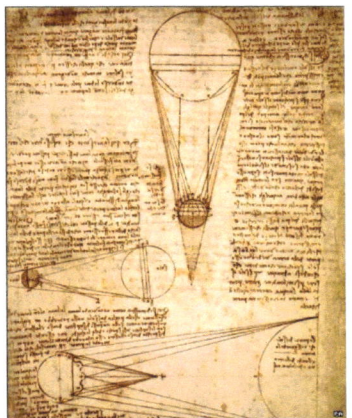

Leonardo Davinci's Invention Notebook

My Great Invention 1

My Great Invention 2

My Great Invention 3

Peanut Products

Affectionately known as the 'Plant Doctor,' George Washington Carver (1860-1943)'s faith was a key component in his inventions. As a science enthusiast, he once asked God to tell him the secrets of the universe, but instead God pointed him to something much smaller – the peanut.

The secrets Carver discovered led to the invention of hundreds of new discoveries including – peanut butter, cosmetics, paint, oil, marble, plywood, and even the dye used in Crayola Crayons. All of his inventions Carver humbly attributes to the Creator having often said, "The Lord has guided me," and "without my Savior, I am nothing."

Sang Ki Lee, Patent Attorney and Attorney at Law, focuses on intellectual property law.

Prior to joining Dream Law, he worked as a patent attorney at SoftBee in Chicago, where he has been involved in helping small businesses and individual inventors for patents, trademarks and copyright and other intellectual property areas.

Sang Lee has extensive experience and expertise in intellectual property rights that require complex, scientific, computer and engineering knowledge. Prior to becoming a lawyer, he worked as a senior researcher and engineer at Panasonic, a global electronics company headquartered in Japan, for 10 years. He also worked as a senior engineer and computer programmer at Landauer, Inc., a global medical device company headquartered in Glenwood, IL, for 10 years. Sang Lee is also an inventor with the patent on credit card processing system at the pump (U.S. Patent No. 5,889,676, granted in 1999)

Patent Attorney, Admitted to the U.S. Patent and Trademark Office
Attorney at Law, Admitted to the State of Illinois
Attorney at Law, Admitted to the United States Tax Court
Attorney at Law, Admitted to Federal Court, Northern District of Illinois
Senior Researcher at Landauer, Inc.
Senior Engineer at Panasonic, Inc.

Loyola University Chicago, J.D.
The University of Illinois, MS in Computer Science
The University of Illinois, BS in Computer Science
Seoul National University, Pre-Dental Program

Email Address: sang@dream-law.com
Phone#: 1-847-357-1358 ext 301   **Fax#: 1-847-357-1359**